LEAKAGE IN NANOMETER CMOS TECHNOLOGIES

SERIES ON INTEGRATED CIRCUITS AND SYSTEMS

Anantha Chandrakasan, Editor
Massachusetts Institute of Technology
Cambridge, Massachusetts, USA

Published books in the series:

A Practical Guide for SystemVerilog Assertions
Srikanth Vijayaraghavan and Meyyappan Ramanathan
2005, ISBN 0-387-26049-8

Statistical Analysis and Optimization for VLSI: Timing and Power
Ashish Srivastava, Dennis Sylvester and David Blaauw
2005, ISBN 0-387-25738-1

Leakage in Nanometer CMOS Technologies
Siva G. Narendra and Anantha Chandrakasan
2005, ISBN 0-387-25737-3

LEAKAGE IN NANOMETER CMOS TECHNOLOGIES

Siva G. Narendra
Tyfone, Inc.

Anantha Chandrakasan
Massachusetts Institute of Technology

 Springer

Library of Congress Cataloging-in-Publication Data

A C.I.P. Catalogue record for this book is available
from the Library of Congress.

ISBN-13: 978-1-4419-3826-8   ISBN-10: 0-387-28133-9 (e-book)

ISBN-13: 978-0-387-28133-9

Printed on acid-free paper.

springeronline.com

Contents

Preface

Scaling transistors into the nanometer regime has resulted in a dramatic increase in MOS leakage (i.e., off-state) current. Threshold voltages of transistors have scaled to maintain performance at reduced power supply voltages. Leakage current has become a major portion of the total power consumption, and in many scaled technologies leakage contributes 30-50% of the overall power consumption under nominal operating conditions. Leakage is important in a variety of different contexts. For example, in desktop applications, active leakage power (i.e., leakage power when the processor is computing) is becoming significant compared to switching power. In battery operated systems, standby leakage (i.e., leakage when the processor clock is turned off) dominates as energy is drawn over long idle periods.

Increased transistor leakages not only impact the overall power consumed by a CMOS system, but also reduce the margins available for design due to the strong relationship between process variation and leakage power. It is essential for circuit and system designers to understand the components of leakage, sensitivity of leakage to different design parameters, and leakage mitigation techniques in nanometer technologies. This book provides an in-depth treatment of these issues for researchers and product designers.

This book also provides an understanding of various leakage power sources in nanometer scale MOS transistors. Leakage sources at the MOS transistor level including sub-threshold, gate tunneling, and junction currents will be discussed. Manifestation of these MOS transistor leakage components at the full chip level depends considerably on several aspects including the nature of the circuit block, its state, its application workload, and process/voltage/temperature conditions. The sensitivity of the various MOS leakage current sources at the transistor level to these conditions will be introduced. These leakage currents at the transistor level translate at the system level in various ways and therefore impact the overall system in a diverse manner. For example, transistor leakages manifest differently under normal operation compared to typical testing conditions, such as burn-in testing. Transistor leakages impact power consumption of the system depending on the system state (e.g., active condition vs. standby condition). Active system leakage power can be significantly higher than standby system leakage, due to elevated temperature and the difficulty to trade-off leakage power for performance. The impact of leakage components also depends on the style of circuit and module type (e.g., memory vs. logic).

To deal with transistor leakage, a variety of solutions is required at all levels of design. The solutions include leakage modeling and prediction, transistor modifications, circuit techniques and system modifications. This book provides an in-depth coverage of promising techniques at the transistor, circuit, and architecture levels of abstraction.

The topics discussed in this book include sources of transistor leakage and its impact, state assignment based leakage reduction, power gating techniques, dynamic voltage scaling, body-biasing, use of multiple performance transistors, leakage reduction in memory, impact of process variation on leakage and design margins, active leakage power reduction techniques, and impact of process variation and leakage on testing. Additionally, two case studies will be presented to highlight real world examples that reap the benefits of leakage power reduction solutions. The last chapter of the book will highlight transistor design choices to mitigate the increase in the leakage components as technology continues to scale.

This book would not have been possible without the concerted effort of all its contributing authors. We would like to thank them for their contribution and help with reviewing other chapters to ensure consistency. We would also like to express sincere thanks to non-contributing reviewers – Dinesh Somashekar and Keith Bowman, both of Intel Corporation. I (Siva) would like to recognize the dedicated contribution of my late colleague and friend at Intel Corporation, Brad Bloechel, without whom, lot of the experimental results in the chapters 2, 6, 8, and 9 would not have been possible. He will be missed. Finally, we want to thank our families for their patience and support through the process of compiling this book together.

Siva G. Narendra
Tyfone, Inc.

Anantha Chandrakasan
Massachusetts Institute of Technology

Chapter 1

TAXONOMY OF LEAKAGE: SOURCES, IMPACT, AND SOLUTIONS

1.1 INTRODUCTION

Benefits of CMOS technology scaling in the nanometer regime comes with the disruptive consequence of increasing MOS transistor leakages. This increase in transistor leakages not only impacts the overall power consumption of a CMOS system, but also reduces the allowed design margins due to the strong relationship between process variation and leakage power. Therefore to continue to reap the benefits of technology scaling, it is essential for circuit designer and system architects to understand the sources of leakage, its impact on circuit and system designs, and solutions to minimize the impact of leakage in such designs. To effectively deal with the impact on circuit and system designs due to leakage, designers need to utilize prediction, reduction, adaptation, and administration techniques.

In this chapter, facts on why leakage power sources are becoming increasingly relevant in CMOS systems that use nanometer scale MOS transistors will be clarified. Leakage sources at the MOS transistor level including sub-threshold diffusion current, gate and junction tunneling currents will be discussed.

These leakage currents at the transistor level manifest themselves in various ways and impact the overall system in a diverse manner. For example, transistor leakages manifest themselves differently under normal operation compared to typical testing conditions, such as burn-in testing. Transistor leakages also impact power consumption of the system differently depending on if the system is in active condition compared or standby condition, as will become obvious later. In a given technology generation, impact of leakage components depends on the style of circuit, such as memory or logic. Additionally the impact on the circuit and systems depends on environmental conditions such as process corner, power supply voltage, and temperature of operation. In this chapter we will explain in further detail the diverse impact transistor leakages have on circuit and systems.

To deal such diverse range of impacts due to transistor leakages, a variety of solutions is required at all levels of design. The solutions constitute leakage prediction, reduction, adaptation, and administration techniques.

By the nature of its origin MOS transistor leakage components in nanometer technologies depend exponentially on parameters, such as oxide thickness, channel length, temperature, and supply voltage. Given that leakage power is expected to become or has already become a significant portion of the total power it is imperative to predict system leakage in the presence of variation in transistor and environmental parameters. In this chapter a statistical approach to predict system leakage will also be introduced.

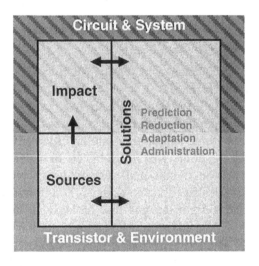

Figure 1-1. Taxonomy of leakage – sources, its impact, and solutions to reduce the impact. Sources originate at the transistor level influenced by the environment. Impact of the sources manifest in various ways at the circuit and system levels. Solutions can be implemented at the transistor, circuit, or system levels. Reduction and adaptation techniques help minimize the impact; prediction helps understand the source and nature of its impact; and administration helps administer and manage an array of solutions.

Reduction of leakage has to be addressed at all levels of the design hierarchy – transistor, circuit, and system. Techniques such as state assignment based leakage reduction, power gating techniques, dynamic voltage scaling, substrate-biasing, use of multiple performance transistors, leakage reduction in memory, active leakage power reduction techniques, and transistor design choices will explained in detailed in dedicated chapters.

Since leakage depends exponentially on several transistor and environmental parameters, it is becoming harder to meet the required system specifications, such as power and performance, over the entire range of these parameters. Adaptation of the system to changes in these parameters helps reduce the impact leakage will have on the design margin. Techniques such as adaptive body bias and adaptive supply voltage will be discussed in detail in latter chapters. Impact of process variation and leakage on testing, and two case studies to highlight real world examples that reap the benefits of leakage reduction solutions will also be discussed.

While administration is not specifically discussed in the book, this is an important aspect of all leakage reduction techniques. It should be clear that there is no single technique that solves all the impacts of leakage. An array of techniques is required and administration of these techniques will be essential to provide effective leakage reduction. Administration includes tools to evaluate the benefits of a technique; tools to productively implement the technique; design, silicon space, and testing resources to put the technique into practice. The two case studies covered will provide examples of aspects of solution administration. The taxonomy of leakage as dealt in this book is illustrated in Figure 1-1.

1.2 SOURCES

To understand the sources of leakage components at the MOS transistor level, it is important to appreciate how transistors and systems that use them have evolved over time to follow Moore's law [1]. MOS transistor based integrated circuits have transformed the world we live in. It is estimated that there are more than 15 billion silicon semiconductor chips currently in use with an additional 500,000 sold each day [2].

The ever shrinking size of the MOS transistors that result in faster, smaller, and cheaper systems have enabled ubiquitous use of these chips. The requisite to continually follow Moore's law and reap its associated benefits is making leakage sources to be more relevant than before. Let us look at the evolution of semiconductor chips that use MOS transistors as the building block.

Among these semiconductor chips, a prevalent component is the high-performance general-purpose microprocessor. Figure 1-2 illustrates the timeline on technology scaling and new high-performance microprocessor architecture introductions in the past three decades. This trend holds in general for other segments of the semiconductor industry as predicted by Moore's law [1]. In 1965, Gordon Moore showed that for any MOS transistor technology there is a minimum cost that maximizes the number of components per integrated circuit. He also showed as transistor dimensions are shrunk (or scaled) from one technology generation to the next, the minimal cost point allows significant increase of the number of components per integrated circuit as shown in Figure 1-3.

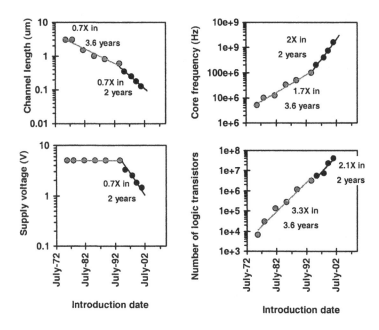

Figure 1-2: Timeline on technology scaling and new microprocessor architecture introduction.

Historically, technology scaling resulted in scaling of vertical and lateral dimensions of the transistor and the associated interconnect structure by 0.7X each generation. This results in delay of the logic gates to be scaled by 0.7X and the integration density of logic gates to be increased by 2X. From the timeline shown in Figure 1-2 it is clear that there were two distinct eras in technology scaling – constant voltage scaling and constant electric field scaling.

Constant voltage scaling era (First two decades): Technology scaling and new architectural introduction in this era happened every 3.6 years. Technology scaling should scale delay by 0.7X translating to 1.4X higher frequency. However, frequency scaled by 1.7X with the additional increase primarily brought about by increase in the number of logic transistors through added circuit and architectural complexity. As it can be seen from Figure 1-2 the number of logic transistors increased by 3.3X in each of the new introductions. Technology scaling itself would have provided only 2X – the additional increase was enabled by increase in die area of about 1.5X every generation [3].

Constant electric field scaling era (Past decade): Technology scaling and new architectural introduction in this era happened every 2 years along with supply voltage (V_{dd}) scaling of 0.7X. As always technology scaling should scale delay by 0.7X translating to 1.4X higher frequency, but frequency increased by 2X in each new introduction. The additional increase in frequency was primarily brought by decrease in logic depth through architectural and circuit design advancements. The number of logic transistors grew only by about 2.1X every generation, which could be achieved without significant increase in die area. Since switching power is proportional to Area x ε/distance x V_{dd} x V_{dd} x F, it increased by (1 x 1/0.7 x 0.7 x 0.7 x 2 =) 1.4X every generation. Although the die size growth is not required for logic transistor integration, it is important to note that the total die area did continue to grow at the rate of 1.5X per generation [3] due to increase amount of integrated memory.

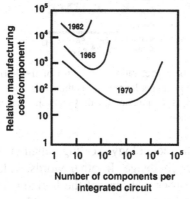

Figure 1-3: Basic form of Moore's law.

In the past decade, technology and new architecture product cycles reduced from 3.6 years to 2 years. From the product development perspective, this requires concurrent engineering in product design, process design, and building of manufacturing supply lines [4]. The past decade also required supply voltage scaling imposed by oxide reliability and the need to slow down the switching power growth rate. The slow down in switching power depends on the magnitude of supply voltage scaling [5]. From the process design stand point supply voltage scaling requires threshold voltage scaling [6, 7] so that the technology scaling can continue to provide 1.4X frequency increase. To prolong the tremendous growth the industry has experienced in the past three decades threshold voltage scaling and concurrent engineering has to continue. These requirements pose several challenges in the coming years including increase in sub-threshold, gate tunneling, and junction tunneling leakage components [7, 8].

1.2.1 Gate tunneling leakage

With scaling of the channel length, maintaining good transistor aspect ratio, by the comparable scaling of gate oxide thickness, junction depth, and depletion depth are important for ideal MOS transistor behavior [7]. The concept of aspect ratio is introduced in Figure 1-4.

$$\text{Device aspect ratio} \approx \frac{L}{\sqrt[3]{T_{ox} \frac{\varepsilon_{si}}{\varepsilon_{ox}} X_j D}}$$

Figure 1-4: MOS transistor aspect ratio is the ratio of the horizontal dimension to the vertical dimension. L is the channel length, T_{ox} is the oxide thickness, D is the measure of the depletion depth, and X_j is the junction depth. Larger the aspect ratio the more ideal the behavior of the MOS transistor.

Unfortunately, with technology scaling, maintaining good transistor aspect ratio has been a challenge. In other words, reduction of the vertical dimensions has been harder than that of the horizontal dimension. With the silicon dioxide gate dielectric thickness approaching scaling limits there is now a rapid increase in gate direct tunneling leakage current [9, 10]. *Figure 1-5* shows the area component of gate leakage current in A/cm^2 versus gate voltage. The oxide thickness limit will be reached approximately when the gate to channel tunneling current becomes equal to the off-state source to drain sub-threshold leakage. This is expected to be ~1 nm physical oxide thickness.

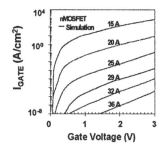

Figure 1-5: Gate leakage versus gate voltage for various oxide thicknesses [11].

Unfortunately, due to quantum mechanical and polysilicon gate depletion effects, both the gate charge and inversion layer charge will be located at a finite distance from the oxide-channel interface with the charge location being a strong function of the bias applied to the gate. The location of the inversion layer charge in the silicon substrate for a transistor with a typical bias when quantum mechanical effects are taken into account is ~1 nm from the oxide-channel interface. This increases the effective oxide thickness by ~0.3 nm (the reduction from 1 nm in silicon to 0.3 nm in oxide is due to difference between the dielectric constants of silicon and silicon dioxide). Taking charge spread on both sides of the interface along with poly depletion, changes the ~1 nm oxide tunneling limit into an effective oxide thickness of ~1.7 nm.

To combat this limit researchers have been exploring several alternatives, including the use of high permittivity gate dielectric, metal gate, novel transistor structures and circuit based techniques [12, 13, 14, 15, 16, 17]. The use of high permittivity gate dielectric will result in thicker and easier to fabricate dielectric for iso-gate oxide capacitance with potential for significant reduction in gate leakage. Identification of a proper high permittivity dielectric material that has good interface states with silicon along with limited gate leakage is in progress [12]. However, it has also been shown that use of high permittivity gate dielectric has limited return [13]. Use of metal gate prevents poly-depletion resulting in a thinner effective gate dielectric. However, identification of dual metal gates to replace the n+ and p+ doped polysilicon is essential to maintain threshold voltage scaling. In addition, novel transistor structures such as self-aligned double gate, FinFET, and tri-gate MOS transistors that promise better transistor aspect ratio [14, 15, 18] are being explored.

1.2.2 Sub-threshold leakage

As discussed earlier, to limit the energy and power increase in future CMOS technology generations supply voltage will have to continually scale. Along with supply voltage scaling, MOS transistor threshold voltage will have to scale to sustain the traditional 30% gate delay reduction. Reduction in threshold voltage results in the increase in sub-threshold leakage current.

To elaborate, in a MOS transistor, when the gate control voltage with reference to the source voltage (V_{gs}) is above the threshold voltage (V_t) the dominant mechanism of drain current is primarily drift based. Drift current in MOS transistors is proportional to $(V_{gs}-V_t)^\alpha$, where $1 \leq \alpha \leq 2$. The drive current of a ON-state MOS transistor which is used to charge or discharge the output capacitor therefore will be proportional to $(V_{dd}-V_t)^\alpha$. This indicates, albeit in an over-simplified manner, that if V_{dd} is reduced there needs to be a corresponding reduction in V_T to maintain the drive current. Now, as the transistor approaches the OFF-state, the V_{gs} goes below V_t. Under this condition the drain current mechanism becomes predominantly diffusion based. Diffusion current, like bipolar transistors, depends exponentially on its control voltage. In other words the drain current changes exponentially with V_{gs} for V_{gs} below V_t. *Figure 1-6* illustrates how reduction in V_t therefore results in larger sub-threshold leakage current. In the illustration, for a sub-threshold swing (S) of 85mV/decade, the sub-threshold leakage current (I_{OFF}) will increase by 10X if the V_t is reduced by 85mV.

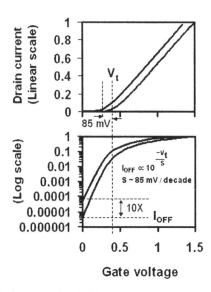

Figure 1-6: Relationship between threshold voltage (V_t) and sub-threshold leakage current (I_{OFF}) for NMOS transistor. Assumes that the source terminal voltage was 0V, so the gate voltage is same as V_{gs}.

Additionally, with technology scaling, the MOS transistor channel length is reduced. As the channel length approaches the source-body and drain-body depletion widths, the charge in the channel due to these parasitic diodes become comparable to the depletion charge due to the MOS gate-body voltage [11], rendering the gate and body terminals to be less effective.

Figure 1-7 shows cross-sectional schematic of long channel and short channel transistors and their corresponding band conduction bands. The band diagram indicates the barrier that majority carriers in the source terminal have to overcome to enter the channel. In a given technology generation, since the source-body and drain-body depletion widths are pre-defined based on the dopings, the rate at which the barrier height increases as a function of distance from the source into the channel is constant. As the band diagram illustrates in Figure 1-7, the finite depletion width of the parasitic diodes do not influence the energy barrier height to be overcome for inversion formation in a long channel transistor.

However, when the channel length is reduced the barrier for the majority carriers to enter the channel also is reduced as indicated in the figure. This results in reduced threshold voltage. In other words, anytime the depletion charge between the source-body and drain-body terminals become a larger fraction of the channel length the threshold voltage reduces. For the same reason in short channel transistor the barrier height, and therefore the threshold voltage are a strong function of the drain voltage. As the figure indicates the barrier reduces as the drain voltage is increased. This barrier lower and drain induced barrier lowering (DIBL) with channel length scaling results in increased sub-threshold leakage currents, apart from the increase in sub-threshold leakage due to V_t reduction required with technology scaling.

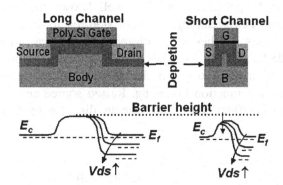

Figure 1-7: Barrier height lowering due to channel length reduction and drain voltage increase in an NMOS field-effect transistor.

It is important to note due to V_t's dependence on channel length and drain voltage in short channel transistors, it is hard to pin point one value V_T for such transistors. For these devices at least $V_{t\text{-LINEAR}}$ (V_t when $V_{ds} \rightarrow 0$) and $V_{t\text{-SATURATION}}$ (V_t when $V_{ds} = V_{dd}$) should be quoted. Also, one of the goals of transistor design is to maximize I_{ON} of a nominal channel length transistor, for a given I_{OFF} of the worst case channel length transistor. The difference between nominal channel length and the worst case channel length arises from channel length spread due to parameter variation. Use of this metric captures the importance of reducing transistor level parameter variation. This metric is also accurate because (i) delay of a critical path is set by the average I_{ON} of transistors in that path and (ii) the leakage power is sum of all the I_{OFF} in the chip therefore will be dominated by the worst case channel length device when considering sub-threshold leakage.

1.2.3 Junction tunneling leakage

To combat the sub-threshold leakage increase due to barrier reduction we need a MOS transistor with good aspect ratio. The challenge of dramatic increase in gate direct tunneling leakage with oxide thickness scaling limits was mentioned in Section 1.2.1. Scaling of junction depth to maintain the aspect ratio, in scaled transistors, leads to increase in the transistor series resistance. Therefore this limits how far the junction depth can be reduced. With channel length reduction, it is therefore necessary to increase the channel doping near the source-to-body and drain-to-body junctions to minimize the effect of barrier lowering. This increased doping in the channel edge is sometimes referred to as halo doping.

As the doping near these junctions are increased with scaling, junction tunneling leakage in the channel edge becomes more prevalent due to the emergence of n+ to p+ junctions. It is well known that reverse biased junctions that have heavy dopings on both side results in direct tunneling across these junctions. Furthermore with reducing volume of the transistors the metal silicide being used for source, gate, and drain terminals increase the probability of creating traps in the nearby heavily doped junctions, further increasing the junction tunneling. Raised source and terminals will help minimize this effect since the silicide distance to the junction is increased.

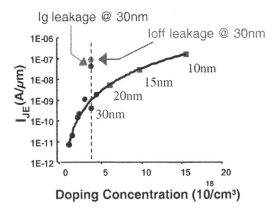

Figure 1-8: Junction edge leakage vs. doping concentration. Circles - data, squares - extrapolated points. Other sources of leakage at 30 nm have been added to the graph. [19]

Figure 1-8 shows the junction edge leakage (I_{JE}) as a function of substrate doping at 25°C and 1V reverse bias. Although the leakages are high (above 1nA/um at 30 nm channel length), they are still a lot less than sub-threshold and gate leakages at 30 nm. For the shorter channel transistors, extrapolating to the 10 nm gate lengths, and assuming a 1.6X doping concentration increase per technology generation, the junction leakage approaches 1uA/um. Note that the data does not comprehend the impact of trap assisted increase in junction leakage.

It is worth noting that one of the scaling limits for the traditional MOS transistor structure arises from the fact that the depletion thicknesses of the drain-to-body and source-to-body junctions are finite. Therefore, there is a minimum channel length below which these junctions will short each other, resulting in direct drain-to-source tunneling. This is predicted to the fundamental limiter for scaling of the traditional MOS transistor structure [20].

1.3 IMPACT

It should be evident by now that in a given technology there is a trade-off between the three leakage sources. Increase in sub-threshold leakage current due to reduction in aspect ratio can be combated by reducing the oxide thickness and/or by increasing the channel doping near the junction – these will result in increase of one or both tunneling leakage currents.

Since different circuit styles may use different types of transistors, the relative importance of the leakage and therefore the optimization of the transistor parameters for each of these transistors may differ. For example, in SRAM circuits compared to logic circuits, it is customary to use longer channel length transistor to minimize impact of random dopant variations. Therefore, SRAM circuits maybe dominated by gate leakage, while logic transistors will be dominated by sub-threshold leakage. Similarly decoupling capacitors that are used to filter power supply noise are long channel MOS transistors. Therefore these transistors will suffer from increase in gate leakage, while sub-threshold and junction leakages have virtually no impact on such decoupling capacitors.

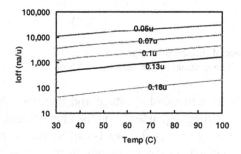

Figure 1-9: Temperature dependence of sub-threshold leakage current per unit um transistor width, under the generational lengths of 0.18 um to 0.05 um.

Sub-threshold leakage current in scaled technologies depends exponentially on temperature (Figure 1-9) since it is a diffusion mode of transport, while the tunneling currents have very weak temperature dependence. So, any temperature variation will significantly affect the sub-threshold leakage current. All leakages components discussed in this section have strong direct exponential dependence on the supply voltage. So any voltage variations will affect the leakage current consumed. Also, more recently the electric field across critical dimensions such as oxide thickness have be increasing for high-performance designs, since thinner oxide are more reliable and can support higher fields. This has resulted in slow down on supply voltage scaling to push the performance envelope. This further aggravates the various leakage current components in high-performance design.

Furthermore, leakage current's strong environmental parameter (voltage, temperature, and process corner) dependencies make the power estimation of such a CMOS system complex. For example, the leakage power consumed by MOS circuit block in active state where the temperature will be higher will be very different from the same circuit in idle state. Note that a nearby circuit block's active state temperature will influence the idle state leakage of another block. The power consumed by a part in burn-in testing, where the temperature and voltage conditions are higher than normal operation, will strongly depend on the leakage sources and their sensitivity to temperature and voltage.

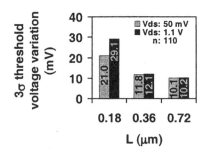

Figure 1-10: Dependence of threshold voltage variation on channel length and drain voltage in 0.18 um channel length generation; n is the number of MOS transistor samples measured.

Additionally, transistor threshold voltage and therefore the sub-threshold leakage power have a strong dependence on the channel length, due the prior explained barrier lowering effects (Figure 1-10). So the power consumed by a circuit block will depend on the variations in the channel lengths of the constituent transistors of that circuit block. With scaling due to improvements in lithography techniques channel length scaling has been more aggressive compared to the ability to reduce oxide thickness and junction depth. This results in worse transistor aspect ratio and barrier lower effects with scaling, leading to increase sub-threshold leakage power dependence on channel length.

1.4 SOLUTIONS

The above mentioned leakage sources and their dependencies on environmental parameters make the impact of the leakage sources at the circuit and system levels quite intricate and diverse. Therefore, there is no *one* solution that will transcend all the negative effects of the transistor leakage sources.

The solution space spans the transistor, circuit, and system levels. They can be further divided in to (i) reduction and adaptation techniques that directly help minimize the impact of leakage and parameter variation, (ii) prediction methodologies that help understand the source and nature of its impact, and (iii) and administration method that help implement and manage an array of solutions.

1.4.1 Reduction and adaptation

The rest of the book will deal with explaining different reduction and adaptation techniques. In Chapter 2, the use of input vectors to minimize the idle leakage of a circuit block with virtually no performance impact is covered. In Chapters 3 and 4, power gating techniques and methodologies are explained. It is worth noting that power gating techniques help address all components of power consumption when a circuit block or a chip is in idle mode, at the expense of some degradation in circuit performance. Chapter 3 covers dynamic voltage scaling based power reduction as well. In Chapter 5 substrate biasing technique that allows electrical modulation of transistor threshold voltage is explained and its usage for power reduction is presented. Chapter 6 covers various adaptive design techniques to minimize impact of parameter variation on power consumption and design margins. Chapter 7 covers memory leakage reduction techniques.

There are two types of leakage power that is of importance (i) active leakage power and (ii) standby leakage power. Active leakage power is defined leakage power consumed by a nanoscale CMOS system when it doing useful work and standby leakage power is consumed when the system is idle. Chapter 8 focuses attention on extending traditionally standby or idle leakage reduction techniques for active leakage reduction. Also, uses of multi-performance transistors for power reduction are described in Chapter 8. Chapter 9 covers impact of leakage and parameter variation on testing. Chapter 12 introduces transistor design in technologies that are dominated by leakage.

1.4.2 Prediction

The present scaling trends have lead to leakage power being as much as 40% of the total power in the 90 nm generation [21]. Under this scenario, it is not only important to be able to reduce leakage power, but also to be able to predict leakage power more accurately. In this section we highlight the importance of including parameter variation in predicting leakage power accurately. Failure to do so will result in gross underestimation or overestimation, both of which are unacceptable.

Due to the wide variation expected in threshold voltage of MOS transistors from die-to-die and within-die during the life time of a process, present leakage current estimation techniques provide lower and upper bounds on the leakage current. The upper and lower bounds are at least an order of magnitude apart and leakage power of most chips lies between the two bounds as shown in [22]. In older technology generations, basing system design on the two leakage current bounds was acceptable since leakage power was a negligible component of the total power. In most systems, the worst case bound is assumed for the design. In technology generations where as much as half of the system power during active mode can be due to leakage, using the worse case bound estimation technique will lead to extremely pessimistic and expensive design solutions. One cannot base the system design on the lower bound since it will lead to overly optimistic and unreliable design solutions. Therefore, it will be crucial to estimate leakage current as accurately as possible. The upper and lower bound estimate equations and measurements are provided in the next part of this section.

The lower bound leakage current estimation of a chip is given as follows,

$$I_{leak-l} = \frac{w_p}{k_p} I_p^o + \frac{w_n}{k_n} I_n^o$$

where, w_p and w_n are the total PMOS and NMOS transistor widths in the chip; k_p and k_n are factors that determine percentage of PMOS and NMOS transistor widths that are in off state; I_p^o and I_n^o are the expected mean leakage currents per unit width of PMOS and NMOS transistors in a particular chip. The mean leakage current is obtained for transistors with mean threshold voltage or channel length. The upper bound leakage current estimation of a chip is related to the transistor leakage as follows,

$$I_{leak-u} = \frac{w_p}{k_p} I^{3\sigma}_{off-p} + \frac{w_n}{k_n} I^{3\sigma}_{off-n}$$

where, $I^{3\sigma}_{off-p}$ and $I^{3\sigma}_{off-n}$ are the worst-case leakage current per unit width of PMOS and NMOS transistors. The worst-case leakage current is obtained for transistors with threshold voltage or channel length 3σ lower than the mean leakage currents per unit width of PMOS and NMOS transistors in a particular chip.

To include the impact of within-die threshold voltage or channel length variation it is necessary to consider the entire range of leakage currents, not just the mean leakage or the worst-case leakage. Let us assume that the within-die threshold voltage or channel length variation follows a normal distribution with respect to transistor width, with μ being the mean and σ being the sigma of the distribution. Let I^o be the leakage of the transistor with the mean threshold voltage or channel length. Then by performing the weighted sum of transistors of different leakage, we can estimate the total leakage of the chip. This is achieved by integrating the threshold voltage or channel length distribution multiplied by the leakage, as shown below.

$$I_{leak} = \frac{I^o w}{k} \frac{1}{\sigma\sqrt{2\pi}} \int_{xmin}^{xmax} e^{\frac{-(x-\mu)^2}{2\sigma^2}} e^{\frac{(\mu-x)}{a}} dx$$

In the above equation, the first exponent estimates the fraction of the total width for the transistor leakage estimated by the second exponent. If the distribution considered within-die is threshold voltage variation then x in the above equation represents threshold voltage and a will be equal to $n\phi_t$ [7]. If the distribution considered is channel length then x in the above equation will represent channel length and a will be equal to λ. λ can be estimated for a technology by measuring the relationship between channel length and transistor leakage. In the rest of this section, we will assume that the distribution of interest is the channel length, since this parameter is used to characterize a technology. Using error function properties, we can simplify the above equation to estimate the leakage of a chip that has both PMOS and NMOS transistors including within-die variation as follows [23],

$$I_{leak-w} = \frac{I^o_p w_p}{k_p} e^{\frac{\sigma_p^2}{2\lambda_p^2}} + \frac{I^o_n w_n}{k_n} e^{\frac{\sigma_n^2}{2\lambda_n^2}}$$

where, w_p and w_n are the total PMOS and NMOS transistor widths in the chip; k_p and k_n are factors that determine percentage of PMOS and NMOS transistor widths that are in off state; I^o_p and I^o_n are the expected mean leakage currents per unit width of PMOS and NMOS transistors in a particular chip; σ_p and σ_n are the standard deviation of channel length variation within a particular chip; λ_p and λ_n are constants that relate channel length of PMOS and NMOS transistors to their corresponding sub-threshold leakages.

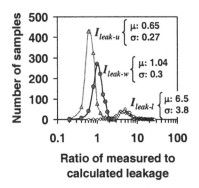

Figure 1-11: Ratio of measured to calculated leakage current ratio distribution for I_{leak-u}, I_{leak-l}, and I_{leak-w} techniques (Sample size: 960).

Measurements in Figure 1-11 indicates the leakage power for most of the samples are underestimated by 6.5X if the lower bound technique is used and overestimated by 1.5X if the upper bound technique is used. The measured-to-calculated leakage ratio for majority of the transistor samples is 1.04 for the technique described in this section. The calculated leakage is within ±20% of the measured leakage for more than 50% of the samples, if the new I_{leak-w} technique is used. Only 11% and 0.2% of the samples fall into this range for the I_{leak-u} and I_{leak-l} techniques respectively. I_{leak-w} technique can be used to predict chip level leakage with better accuracy once transistor level leakage, parameter variation, and total transistor widths are known.

We explained in this section how channel length parameter variation needs to be comprehended to improve the prediction accuracy of a standby leakage power in a system dominated by sub-threshold leakage. Similarly, for active leakage variation in temperature and power supply voltage will have to be comprehended. Also, in general in variation in transistor parameters that significantly modify the tunneling current sources will also have to be comprehended to improve the prediction [24].

1.4.3 Administration

An integrated processing system offering over 200 Giga instructions per second, with 2 billion logic transistors and additionally an order of magnitude more memory transistors, using less than 20 nm physical gate length transistors, operating below 700 mV supply voltage by year 2015 – this is the expected roadmap should the scaling trends continue. Can we achieve this – maybe, maybe not! Nevertheless to attempt at implementing the vision of such a processing system, it is essential that its design comprehend leakage power and parameter variation. Given that there is no global scheme to solve all sources of leakage and its impact, and given that these solutions could transcend across the hierarchy of design from transistors to systems, it is essential to have a comprehensive administration of the different techniques. While this topic is not covered in this book, Chapters 10 and 11 discuss real world examples that highlight administration of leakage power reduction techniques.

The expected evolution of present day CMOS VLSI computational units to nanoscale CMOS VLSI computational units is speculated in Figure 1-12. Essential features include adaptive techniques to reduce design margins, special purpose computation units to improve computational energy efficiency, dense memory choices that enable continued scaling of integrated random access memory, and effective power management schemes that while occupying silicon area enable integration of additional transistors for computation. All of this will be possible if and only if there is cohesive interaction between transistor, circuit, architecture, and platform designers!

Therefore, as the scaling trend continues it will be imperative to develop comprehensive administration standards, since the success of the minimizing the impact of leakage power will require collaboration across several hierarchies of design teams and corporations.

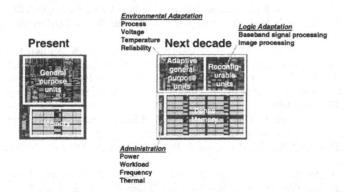

Figure 1-12: Speculated evolution for future nanoscale silicon based CMOS systems.

REFERENCES

[1]   G.E. Moore, "Cramming more components onto integrated circuits," *Electronics*, vol. 38, no. 8, April 19, 1965.

[2]   R. Smolan and J. Erwitt, *One Digital Day – How the Microchip is Changing Our World*, Random House, 1998.

[3]   V. De and S. Borkar, "Technology and Design Challenges for Low Power & High Performance," *Intl. Symp. Low Power Electronics and Design*, pp. 163-168, Aug. 1999.

[4]   K.G. Kempf, "Improving Throughput across the Factory Life-Cycle," *Intel Technology Journal*, Q4, 1998.

[5]   A. Chandrakasan, S. Sheng, and R. W. Brodersen, "Low-Power CMOS Digital design," *IEEE J. Solid-State Circuits*, vol. 27, pp. 473-484, Apr. 1992.

[6]   S. Thompson, P. Packan, and M. Bohr, "MOS Scaling: Transistor Challenges for the 21st Century," *Intel Technology Journal*, Q3, 1998.

[7]   Y. Taur and T. H. Ning, Fundamentals of Modern VLSI Transistors, Cambridge University Press, 1998

[8]   D. Antoniadis and J.E. Chung, "Physics and Technology of Ultra Short Channel MOSFET Transistors," *Intl. Electron transistors Meeting*, pp. 21-24, 1991.

[9]   D.A. Muller, T. Sorsch, S. Moccio, F.H. Baumann, K. Evans-Lutterodt, and G. Timp, "The Electronic Structure at the Atomic Scale of Ultrathin Gate Oxides," *Nature*, vol. 399, pp. 758-761, June 1999.

[10] M. Schulz, "The End of the Road for Silicon," *Nature*, vol. 399, pp. 729-730, June 1999.

[11] S.-H.Lo, D.A. Buchanan, Y. Taur, and W. Wang, *IEEE Electron Transistor Letter*, 1997, p. 209.

[12] C. H. Lee, S. J. Lee, T. S. Jeon, W. P. Bai, Y. Sensaki, D. Roberts, and D. L. Kwong, "Ultra Thin ZrO(2) and Zr(27)Si(10)O(63) Gate Dielectrics Directly Prepared on Si-Substrate by Rapid Thermal Processing," *SRC Techcon*, pp. 46, Sep. 2000.

[13] N. R. Mohapatra, M. P. Desai, S. Narendra, and V. R. Rao, "The effect of high-K gate dielectrics on deep submicrometer CMOS transistor and circuit performance," *IEEE Transactions on Electron Transistors*, vol. 49, pp. 826-831, May 2002.

[14] J. Lee, G. Tarachi, A. Wei, T. A. Langdo, E. A. Fitzgerald, D. Antoniadis, "Super self-aligned double-gate (SSDG) MOSFETs utilizing oxidation rate difference and selective epitaxy," *Intl. Electron Transistors Meeting*, pp. 71-74, 1999.

[15] R. Chau, B. Boyanov, B. Doyle, M. Doczy, S. Datta, S. Hareland, B. Jin, J. Kavalieros, M. Metz, "Silicon nano-transistors for logic applications," *Physica E, Low-dimensional Systems and Nanostructures*, Vol. 19, Issues 1-2, pp. 1-5, July 2003.

[16] I. Kohno, T. Sano, N. Katoh, and K. Yano, "Threshold Canceling Logic (TCL): A Post-CMOS Logic Family Scalable Down to 0.02 mm," *Intl. Solid-State Circuits Conf.*, pp. 218-219, 2000.

[17] T. Kuroda, T. Fujita, S, Mita, T. Nagamatsu, S. Yoshioka, K. Suzuki, F. Sano, M. Norishima, M. Murota, M. Kako, M. Kinugawa, M. Kakumu, and T. Sakurai, "A 0.9-V, 150-MHz, 10-mW, 4-mm2, 2-D Discrete Cosine Transform core Processor with Variable Threshold-Voltage (VT) Scheme," *IEEE J. Solid-State Circuits*, vol. 31, pp. 1770-1779, Nov. 1996.

[18] X. Huang, W-C. Lee, C. Kuo, D. Hisamoto, L. Chang, J. Kedzierski, E. Anderson, H. Takeuchi, Y-K. Choi, K. Asano, V. Subramanian, T-J. King, J. Bokor, C. Hu, "Sub 50-nm FinFET: PMOS," *IEDM Technical Digest*, Washington, DC, pp. 67-70, December 5-8, 1999.

[19] Doyle, B., Arghavani, R., Barlage, D., Datta, S., Doczy, M., Kavalieros, J., Murthy, A., and Chau, R., "Transistor Elements for 30nm Physical Gate Lengths and Beyond." *Intel*

Technology Journal. http://developer.intel.com/technology/itj/2002/volume06issue02/ (May 2002).

[20] B. Hoeneisen, and C.A. Mead, "Fundamental Limitations in Microelectronics I: MOS Technology," *Solid-State Electronics*, vol. 15, pp. 819-829, July 1972.

[21] A. Grove, http://www.intel.com/pressroom/archive/speeches/grove_ 20021210.pdf, *IEDM 2002 Keynote Luncheon Speech.*

[22] A. Keshavarzi, S. Ma, S. Narendra, B. Bloechel, K. Mistry, T. Ghani, S. Borkar, and V. De, "Effectiveness of reverse body bias for leakage control, in scaled dual Vt CMOS ICs," *Intl. Symp. Low Power Electronics and Design*, pp. 207-212, Aug. 2001.

[23] S. Narendra, V. De, S. Borkar, D. Antoniadis, and A. Chandrakasan, "Full-chip subthreshold leakage power prediction and reduction techniques for sub-0.18-um CMOS," *IEEE Journal of Solid-State Circuits*, vol. 39, pp. 501-510, Sept. 2004.

[24] Dongwoo Lee, Wesley Kwong, David Blaauw, Dennis Sylvester, "Analysis and Minimization Techniques for Total Leakage Considering Gate Oxide Leakage," *ACM/IEEE Design Automation Conference (DAC)*, pp. 175-186, June 2003.

Chapter 2

LEAKAGE DEPENDENCE ON INPUT VECTOR

Siva Narendra[§], Yibin Ye[¶], Shekar Borkar[¶], Vivek De[¶], and Anantha Chandrakasan[*]

[§]*Tyfone, Inc., USA*, [¶]*Intel Corp., USA, and *Massachusetts Institute of Technology, USA*

2.1 INTRODUCTION

As described earlier to limit the energy and power increase in future CMOS technology generations, the supply voltage (V_{dd}) will have to continually scale. The amount of energy reduction depends on the magnitude of V_{dd} scaling. Along with V_{dd} scaling, the threshold voltage (V_t) of MOS transistors will have to scale to sustain the traditional 30% gate delay reduction. These V_{dd} and V_t scaling requirements pose several technology and circuit design challenges. In this chapter the term leakage refers to sub-threshold leakage, unless otherwise explicitly mentioned.

One of challenge with technology scaling is the rapid increase in sub-threshold leakage power due to V_t reduction. Should the present scaling trend continue it is expected that the sub-threshold leakage power will become a considerable constituent of the total dissipated power. In such a system it becomes crucial to identify techniques to reduce this leakage power component. It has been shown previously that the stacking of two off transistors has significantly reduced sub-threshold leakage compared to a single off transistor. The stack effect can therefore be used not jus for leakage reduction by forcing stacks, but also using natural stacks that existing in logic gates. Natural stacks can be realized by loading an appropriate primary input vector such that it propagates to maximize the total channel width of stacked transistors that are *OFF*.

In this chapter we present a model that predicts the stack effect factor, which is defined as the ratio of the leakage current in one off transistor to the leakage current in a stack of two off transistors [1]. Model derivation based on transistor fundamentals and verification of the model through statistical transistor measurements from 0.18 μm and 0.13 μm technology generations are presented. The scaling nature of the stack effect leakage reduction factor is also discussed. The derived model for leakage reduction depends on fundamental transistor parameter. This makes the model viable to predict potential leakage savings using stack effect techniques in future transistors.

There are number of solutions including reverse body bias, power gating, and multi-performance transistors can be used to reduce power during standby mode. All of these will be discussed in detailed in the later chapters.

In this chapter after the introduction of stack effect, we will review a new standby leakage control scheme which exploits the large reduction in leakage current achievable by simultaneously turning *OFF* more than one transistor in NMOS or PMOS stacks. Usually, a large circuit block consists of a significant number of logic gates where transistor stacks already exist, such as the PMOS stack in NOR or NMOS stacks in NAND gates.

This first solution, using stack effect in natural stacks that already exists, enables effective leakage reduction during standby mode by installing a vector at the inputs of the circuit block so as to maximize the number of PMOS and NMOS stack with more than one *OFF* transistor. In contrast to the other leakage reduction techniques this scheme offers leakage reduction with minimal overheads in area, power, and technology requirements. Extensive circuit simulations of a sample circuit block to (*a*) elucidate the dynamics of leakage reduction using transistor stacks, (*b*) influence on overall leakage power reduction of the circuit block during both active and standby modes of operation, and (*c*) determine the standby leakage reductions due to the use of natural stacks will be discussed [2].

Another solution to the problem of ever-increasing leakage is to force a non-stack transistor to a stack of two transistors without affecting the input load. By ensuring iso-input load, the previous gate's delay and the switching power will remain unchanged. Logic gates after stack forcing will reduce leakage power, but incur a delay penalty, similar to replacing a low-V_t transistor with a high-V_t transistor in a dual-V_t design. In a dual-V_t design the low-V_t transistors are used in performance critical paths and the high-V_t transistors in the rest. Further details of dual-V_t design technique will be described in Chapter 8 under multi-performance transistors.

Usually a significant fraction of the transistors can be high-V_t or forced-stack since a large number of the paths are non-critical. This will reduce the overall leakage power of the chip without impacting operating clock frequency. In this chapter we discuss the stack forcing method to reduce leakage in paths that are not performance critical. This stack forcing technique can be either used in conjunction with dual-V_t or can be used to reduce the leakage in a single-V_t design.

Although it is not covered in depth in this chapter, it should be pointed out that vector dependent leakage behavior can not only be used to reduce standby sub-threshold current, but also total standby leakage current in the presence of tunneling sources. The current of transistor due to just the gate leakage is more when a transistor is *ON* compared to *OFF*, due to larger area. The gate leakage area of a transistor that is *OFF* is usually just the drain-gate overlap area, while in the case of a transistor that is *ON* it usually includes the drain-gate overlap, source-gate overlap, and channel areas. This is reverse of sub-threshold leakage current, therefore understanding of the relative contribution of the different leakage currents and proper methodology to identify the leakage minimizing input vector is critical [3]. Having said that, it is also necessary to realize under most conditions for logic circuits sub-threshold leakage will be a more dominant component.

2.2 STACK EFFECT

To reiterate, should the present scaling trend continue it is expected that the sub-threshold leakage power will become as much as 50% of the total power in the 0.09 μm generation [4]. Under this scenario, it is not only important to be able to predict sub-threshold leakage power more accurately as discussed in the previous section, it becomes crucial to identify techniques to reduce this leakage power component. It has been shown previously that the stacking of two *OFF* transistors has significantly reduced sub-threshold leakage compared to a single *OFF* transistor [2, 5, 6]. This concept of stack effect is illustrated in Figure 2-1.

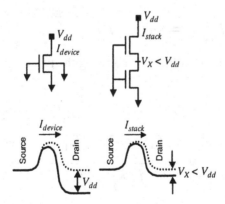

Figure 2-1. Leakage difference between a single *OFF* transistor and a stack of two *OFF* transistors. As illustrated by the energy band diagram, the barrier height is modulated to be higher for the two-stack due to smaller drain-to-source voltage resulting in reduced leakage.

In this section, a model is derived that predicts the stack effect factor, which is defined as the ratio of the leakage current in one *OFF* transistor to the leakage current in a stack of two *OFF* transistors. Model derivation based on transistor fundamentals and verification of the model through statistical transistor measurements from 0.18 μm and 0.13 μm technology generations are presented. The scaling nature of the stack effect leakage reduction factor is also discussed.

Let I_1 be the leakage of a single transistor of unit width in *OFF* state with its $V_{gs} = V_{bs} = 0$ V and $V_{ds} = V_{dd}$. If the gate-drive, body bias, and drain-to-source voltages reduce by ΔV_g, ΔV_b, and ΔV_d respectively from the above-mentioned conditions, the leakage will reduce to,

$$I'_1 = I_1 \ \ 10^{-\frac{1}{S}\left[\Delta V_g + \lambda_d \Delta V_d + k_\gamma \Delta V_b\right]}$$

where S is the sub-threshold swing, λ_d is the drain-induced barrier lowering (DIBL) factor, and k_γ is the body effect coefficient. The above equation assumes that the resulting $V_{ds} > 3kT/q$ [7]. For a two-transistor stack shown in Figure 2-2 a steady state condition will be reached when the intermediate node voltage V_{int} approaches V_x such that the leakage currents in the upper and lower transistors are equal. Under this condition, the leakage currents in the upper and lower transistors can be expressed as,

$$I_{stack-u} = w_u I_1 \ 10^{\frac{-(1+\lambda_d+k_\gamma)V_x}{S}}$$

$$I_{stack-l} = w_l I_1 \ 10^{\frac{-\lambda_d(V_{dd}-V_x)}{S}}$$

and the intermediate node voltage by equating the two current can be derived to be,

$$V_x = \frac{\lambda_d V_{dd} + S\log\dfrac{w_u}{w_l}}{1 + k_\gamma + 2\lambda_d}$$

For short channel transistors the body terminal's control on the channel is negligible compared to gate and drain terminals, implying $k_\gamma \ll 1 + 2\lambda_d$.

Hence, the steady state value, V_x, of the intermediate node voltage can be approximated as,

$$V_x \approx \frac{\lambda_d V_{dd} + S \log \frac{w_u}{w_l}}{1+2\lambda_d}$$

Substituting V_x in either $I_{stack\text{-}u}$ or $I_{stack\text{-}l}$ will yield the leakage current in a two-stack given by,

$$I_{stack} = w_u^{\alpha} w_l^{1-\alpha} I_1 \; 10^{\frac{-\lambda_d V_{dd}}{S}(1-\alpha)}$$

$$where \quad \alpha \approx \frac{\lambda_d}{1+2\lambda_d}$$

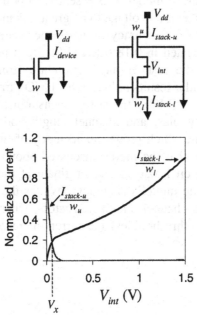

Figure 2-2. Load line analysis showing the leakage reduction in a two-stack.

The leakage reduction achievable in a two-stack comprising of transistors with widths w_u and w_l compared to a single transistor of width w is given by,

$$X = \frac{I_{device}}{I_{stack}} = \frac{w}{w_u^{\alpha} w_l^{1-\alpha}} 10^{\frac{\lambda_d V_{dd}}{S}(1-\alpha)}$$

$$= 10^{\frac{\lambda_d V_{dd}}{S}(1-\alpha)} \quad \text{when} \quad w_u = w_l = w$$

The stack effect factor, when $w_u = w_l = w$, can be rewritten as,

$$X = 10^{\frac{\lambda_d V_{dd}}{S}\left(\frac{1+\lambda_d}{1+2\lambda_d}\right)} = 10^U$$

where U is the universal two-stack exponent which depends only on the process parameters, λ_d and S, and the design parameter, V_{dd}. Once these parameters are known, the reduction in leakage due to a two-stack can be determined from the above model. It is essential to point out that the model assumes the intermediate node voltage to be greater than $3kT/q$.

To confirm the model's accuracy we performed transistor measurements on test structures fabricated in 0.18 µm and 0.13 µm process technologies. Results discussed in the rest of the section are from NMOS transistor measurements, but similar results hold true for PMOS transistors as well.

Figure 2-3 shows NMOS transistor measurements under different temperature, V_{dd}, body bias, and channel length conditions for 0.18-µm technology generations, which prove the accuracy of the theoretical model. It is important to note that the model discussed above doesn't include the impact of diode junction leakages that originate at the intermediate stack node. In Figure 2-3, the model's accuracy deviates the most under reverse body bias for nominal channel length transistors, where the ratio of diode junction leakage to sub-threshold leakage current increases.

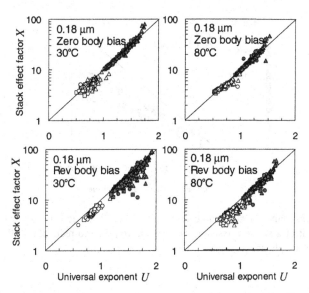

Figure 2-3. Measurement results showing the relationship between stack effect factor X for a two-stack to the universal exponent U. Lines indicate the relationship as per the analytical model and symbols are from measurement results. White symbols are for nominal channel transistors and gray symbols are for transistors smaller than the nominal channel length. Triangle, circle, and square symbols are for V_{dd} of 1.5, 1.2, and 1.1 V respectively. Zero body bias is when the body-to-source diode of the transistor closet to the power supply is zero biased and reverse body bias is when the diode is reverse biased by 0.5 V.

It is known that the stack effect factor strongly depends on λ_d as suggested by the model. In addition, a decrease in the channel length (L) will increase λ_d in a given technology [8]. So, any increase in the leakage of a single transistor due to decrease in L will not increase leakage of a two-stack at the same rate. This is illustrated in Figure 2-4 where increase in two-stack leakage is at a slower rate than that of a single transistor. Therefore, variation in L will result in smaller effective threshold voltage variation for a two-stack compared to a single transistor. Figure 2-5 illustrates the average stack effect factor for the nominal channel transistors in both 0.18 μm and 0.13 μm technology generations obtained from both the measurements and the model. The increase in stack effect factor at a given V_{dd} with technology scaling is attributed to increase in λ_d, which is predicted by the analytical model. The higher stack effect factor for the low-V_t transistor in 0.13 μm technology generation is due to the same effect.

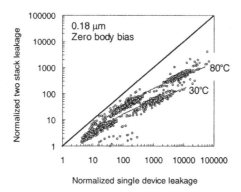

Figure 2-4. Measurement results indicate a slower rate of increase in leakage of two-stack compared to that of a single transistor. This should translate to reduction in the variation of effective threshold voltage.

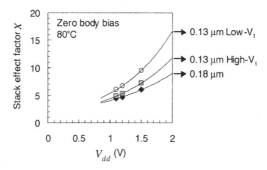

Figure 2-5. Nominal channel length transistor measurement results showing stack effect factor across two technology generations. The increase in stack effect factor is attributed to worsening of short channel effect, λ_d, which is predicted by the analytical model. The higher stack effect factor for the low-V_t transistor in 0.13 μm technology generation is attributed to the same reason. Lines are from analytical model and symbols are from measurement.

In 0.13-μm generation, the low-V_t transistor will dominate chip leakage. Figure 2-6 shows the scaling of stack effect from a 0.18 μm transistor to a 0.13 μm low-V_t transistor based on transistor measurements under different V_{dd} scaling scenarios. Since λ_d is expected to increase due to worsening transistor aspect ratio and since V_{dd} scaling will slow down due to related challenges [9], stack effect leakage reduction factor is expected to increase with technology scaling. The predicted scaling of stack effect factor from 0.18 μm to 0.06 μm is depicted in Figure 2-7.

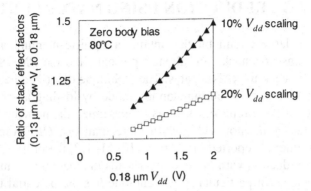

Figure 2-6. Nominal channel length transistor measurement results indicating the scaling of stack effect factor from 0.18 μm to 0.13 μm low-V_t under different V_{dd} scaling conditions. The low-V_t transistor will dominate leakage in 0.13 μm technology, so the comparison is made with the low-V_t transistor.

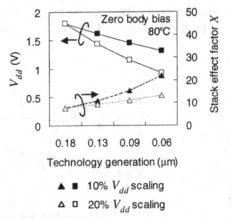

Figure 2-7. Prediction in the scaling of stack effect factor for two V_{dd} scaling scenarios in nominal channel length transistors. V_{dd} for 0.18 μm is assumed to be 1.8 V.

This scaling nature of stack effect factor makes it a powerful technique for leakage reduction in future technologies. In the next sections, we describe a circuit technique for taking advantage of stack effect to reduce leakage at a functional block level. In the first case, the natural stacks present in circuit blocks are used to reduce leakage in standby state, by loading appropriate input vectors to maximize amount of transistor width in stack mode. In the next case, forced stacks are used to minimize leakage of transistors in non-performance critical paths.

2.3 LEAKAGE REDUCTION USING NATURAL STACKS

Typically, a large circuit block contains a significant number of logic gates where transistor stacks are already present, like the PMOS stack in NOR or NMOS stack in NAND gates. The technique described here enables effective leakage reduction during standby mode by loading a vector at the primary inputs of the circuit block so as to maximize the number of PMOS and NMOS stack transistor widths with more than one *OFF* transistor. In contrast to techniques reported in the past [10, 11, 12], the proposed scheme offers leakage reduction with minimal overheads in area, power, and process technology change. In particular, this technique has the potential to replace the need for a high-V_t transistor for standby leakage.

Extensive results from circuit simulations of individual logic gates and a 32-bit static CMOS adder, designed in a 0.1 μm, is discussed to elucidate the dynamics of leakage reduction due to transistor stacks, examine its influence on the overall leakage power of the adder during both active and standby modes of operation, and determine the standby leakage reductions yielded by application of the new leakage control technique. Two different V_t values were considered throughout the analysis. The low-V_t is 100 mV smaller than the high-V_t.

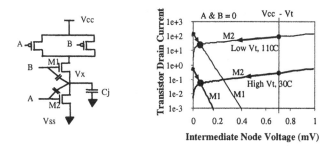

Figure 2-8. 2 NMOS stack in a NAND gate and DC solution for intermediate node voltage.

A 2-input NAND gate is used to illustrate the dynamics of leakage reduction in 2-transistor stacks with both transistors *OFF*, as shown in Figure 2-8. From the DC solution of NMOS sub-threshold current characteristics, shown in Figure 2-8, it is clear that the leakage current through a 2-transistor stack is approximately an order of magnitude smaller than the leakage of a single transistor. This reduction in leakage is can be viewed to come about due to negative gate-to-source biasing and body-effect induced V_t increase in M1, or reduced drain-to-source voltage in M2 which causes its V_t to increase, as the voltage V_x at the intermediate node converges to ~100 mV. Thus, as shown in Figure 2-9, smaller amounts of leakage reduction are obtained at higher temperatures due to larger sub-threshold swing. For 3- or 4-transistor stacks, the leakage reduction is found to be 2-3X larger in both NMOS and PMOS, as illustrated in Figure 2-10.

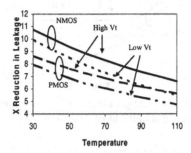

Figure 2-9. Leakage reduction in 2 NMOS and 2 PMOS stacks at different temperatures and different target threshold voltages, from simulations.

	High Vt	Low Vt
2 NMOS	10.7X	9.96X
3 NMOS	21.1X	18.8X
4 NMOS	31.5X	26.7X
2 PMOS	8.6X	7.9X
3 PMOS	16.1X	13.7X
4 PMOS	23.1X	18.7X

Figure 2-10. Leakage current reduction in multiple stacked transistors.

It is essential that we point out an anomaly – according to Figure 2-10, the simulation results show that low-V_t transistors have lower leakage reduction compared to high-V_t transistors. This is contradictory to the measurements and the model derived in the previous section. Low-V_t transistors have larger DIBL therefore should have larger leakage reduction due to stack effect as per the measurements and model. The simulation results due to the models used do not predict the expected behavior of leakage reduction due to stack effect when the V_t is lowered.

Generally speaking, this should be a note of caution to the reader, do not always believe the simulations without proper validation! Absolute values of measured results will probably be different from the simulation results described in this section. It is also important to keep in mind, that measured results will always have a statistical spread of values instead of a single value due to the impact of process variation on leakage, as shown in the previous section. Other than the mentioned threshold voltage related anomaly the simulation result's ability to quantify the benefit of natural stacks for leakage reduction presented in this section holds.

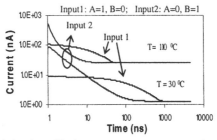

Figure 2-11. Transient behavior of leakage current convergence time constant in a 2 NMOS stack under different temperature and initial input conditions.

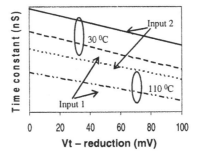

Figure 2-12. Dependence of leakage convergence time constant of stack leakage on threshold voltage, temperature, and initial input conditions.

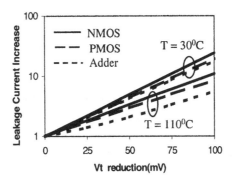

Figure 2-13. Leakage current increase with threshold voltage reduction at the transistor and adder block levels.

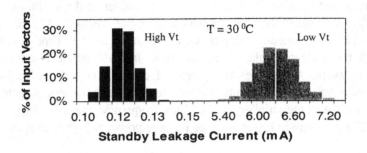

Figure 2-14. Distribution of standby leakage current in the 32-bit adder for a large number of random input vectors.

Back to the simulated data, the time required for the leakage current in transistor stacks to converge to its final value is dictated by the rate of charging or discharging of the capacitance at the intermediate node by the sub-threshold drain current of M1 or M2. This time constant as shown in Figure 2-11 is, therefore, determined by drain-body junction and gate-overlap capacitances per unit width, the input conditions immediately before the stack transistors are turned *OFF*, and transistor sub-threshold leakage current, which depends strongly on temperature and V_t. Therefore, the convergence rate of leakage current in transistor stacks increases rapidly with V_t reduction and temperature increase, as shown in Figure 2-11 and Figure 2-12. For Low-V_t transistors in the 0.1 μm technology, this time constant in 2-NMOS stacks at 110°C ranges from 5-50 ns depending on input conditions before both transistors are turned *OFF*.

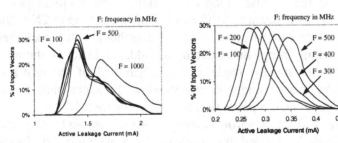

Figure 2-15. Distribution of active leakage current in the 32-bit adder with low-Vt transistors (left) and high-Vt transistors (right) at different frequencies.

Increase in the active and standby leakage of the 32-bit static CMOS Kogge-Stone adder with V_t-reduction, as shown in Figure 2-13, is smaller than that in individual transistors, due to the presence of a significant number of transistor stacks in the design. The standby leakage power varies by 30%-40%, depending on the input vector, as shown in Figure 2-14, which determines the number of transistor stacks in the design with more than one

OFF transistor. Figure 2-15 shows that the adder leakage during active operation is dictated by the sequence of input vectors as well as the operating clock frequency. Magnitude of the stack leakage time constant at elevated temperatures relative to the time interval between consecutive switching events determines the extent of convergence of the leakage to steady-state value. As a result, the active leakage corresponding to each input vector becomes higher as the clock frequency increases from 100 to 1000 MHz resulting in larger average leakage power at higher frequencies.

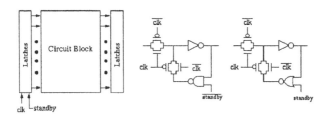

Figure 2-16. Implementation of the standby leakage control using natural stacks through input vector activation.

Figure 2-16 shows an implementation of the new leakage reduction technique where a standby control signal, derived from the clock gating signal, is used to generate and store a predetermined vector in the static input latches of the adder during standby mode so as to maximize the number of NMOS and PMOS stacks with more than one OFF transistor. Since the desired input vector for leakage minimization is encoded by using a NAND or NOR gate in the feedback loop of the static latch, minimal penalty is incurred in adder performance. As shown in Figure 2-17, up to 2X reduction in standby leakage can be achieved by this technique. In order that the additional switching energy dissipated by the adder and latches, during entry into and exit from "standby mode", be less than 10% of the total leakage energy saved by this technique during standby, the adder must remain in standby mode for at least 5 µs, as summarized in Figure 2-18.

A standby leakage control technique, which exploits the leakage reduction offered by natural transistor stacks, was presented. Based on simulation results that showed up to 10X leakage reduction at gate level resulted in up to 2X reduction in standby leakage power. By using natural stacks this can be achieved with minimal overheads in area, power, and process technology change. We also elucidated the dynamics of leakage reduction due to transistor stacks, and its influence on overall leakage power of large circuits. Since with technology scaling the leakage reduction due to stack effect is expected to increase as described in the previous section, this technique will become more effective. Additionally, the time constant for leakage convergence depends on the sub-threshold leakage current itself, so with scaling this time constant will reduce rapidly due to exponential increase in sub-threshold leakage.

High Vt	Avg. Worst	% Reduction 35.4% 60.7%
Low Vt	Avg. Worst	33.3% 56.5%

Figure 2-17. Adder leakage reduction using the best input vector activation compared to the average and worst case standby leakage causing input vectors.

	High Vt	Low Vt
Savings	2.2uA	0.0384mA
Overhead	1.64 nJ	1.84 nJ
Min. time in standby	84 uS	5.4 uS

Figure 2-18. Standby leakage power savings and the minimum time required in standby mode.

2.4 LEAKAGE REDUCTION USING FORCED STACKS

As shown earlier, stacking of two transistors that are *OFF* has significantly reduced leakage compared to a single *OFF* transistor. However due to the iso-input load requirement and due to stacking of transistors, the drive current of a forced-stack gate will be lower resulting in increased delay. So, stack forcing can be used only for paths that are non-critical, just like using high-V_t transistors in a dual-V_t design [13, 14]. Forced-stack gates will have slower output edge rate similar to gates with high-V_t transistors. Figure 9 illustrates the use of techniques that provide delay-leakage trade-off. As demonstrated in the figure, paths that are faster than required can be slowed down which will result in leakage savings. Such trade-offs are valid only if the resulting path still meets the target delay. Figure 2-19 shows the delay-leakage trade-off due to n-stack forcing of an inverter with fan-out of 1 under iso-input load conditions in a dual-V_t 0.13 μm technology [15].

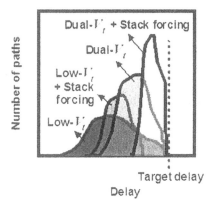

Figure 2-19. Stack forcing and dual-V_t can reduce leakage of gates in paths that are faster than required.

By properly employing forced-stack one can reduce standby and active leakage of non-critical paths even if a dual-V_t process is not available. This method can also be used in conjunction with dual-V_t. Stack forcing provides wider coverage in the delay-leakage trade-off space as illustrated in Figure 2-20.

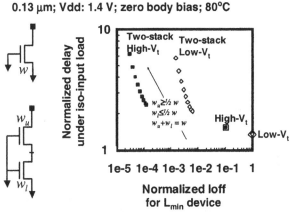

Figure 2-20. Simulation result showing the delay-leakage trade-off that can be achieved by stack forcing technique under iso-input load conditions. Iso-input load is achieved by making the gate area after stack forcing identical to before stack forcing. Several such conditions are possible, which enhances delay-leakage trade-off possible by stack forcing. The two-stack condition for a given V_t with the least delay is for $w_u=w_l=\frac{1}{2}w$. This trade-off can be used with or without high-V_t transistors. The simulation anomaly described in Section 2.3 for Figure 2-10 is evident here as well.

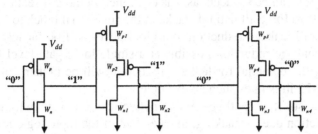

Figure 2-21. A sample path where natural stack is used to reduce standby leakage by applying a predetermined vector during standby. No delay penalty is incurred with this technique.

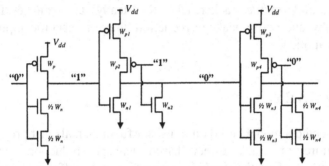

Figure 2-22. Using stack-forcing technique the number of logic gates in stack mode can be increased. This will enable further leakage reduction in standby mode. Increase in delay under normal mode of operation will be incurred.

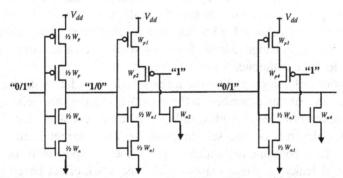

Figure 2-23. If a gate can have its input as either "0" or "1" and still force stack effect then that gate will have reduced active leakage. The more the number of inputs that can be either "0" or "1" the higher the probability that stack effect will reduce active leakage.

Functional blocks have naturally stacked gates such as NAND, NOR, or other complex gates. By maximizing the number of natural stacks in *OFF* state during standby by setting proper input vectors, the standby leakage of functional block can be reduced, as was explained in the last section. Since it is not possible to force all natural stacks in the functional block to be in *OFF* state the overall leakage reduction at a block level will be far less than the stack effect leakage reduction possible at a single logic gate level [2]. With stack forcing the potential for leakage reduction will be higher. Figure 2-21 and Figure 2-22 illustrates such an example.

Forcing a stack in both n- and p-networks of a gate will guarantee leakage reduction due to stacking, independent of the input logic level. Such an example is shown in Figure 2-23. To reiterate, stack forcing can be applied to paths only if increase in delay due to stacking does not violate timing requirements. Gates that can force stack effect independent of its input vectors will automatically go into leakage reduction mode when the intermediate node of the stack reaches the steady state voltage. This will boost standby and active leakage reduction since no specific input vector needs to be applied.

2.5 SUMMARY

We presented a model based on transistor fundamentals that predicted the scaling nature of stack effect based leakage reduction. Transistor measurements verified the model's accuracy across different temperature, channel length, body bias, supply voltage, and process technology.

A standby leakage control technique, which exploits the leakage reduction offered by natural transistor stacks, was presented. Based on simulation results that showed up to 10X leakage reduction at gate level resulted in up to 2X reduction in standby leakage power. By using natural stacks this can be achieved with minimal overheads in area, power, and process technology change. Modes for using stack forcing to reduce standby and active leakage components were discussed.

Since with technology scaling the leakage reduction due to stack effect is expected to increase as described in the previous section, this technique will become more effective. Additionally, the time constant for leakage convergence depends on the sub-threshold leakage current itself, so with scaling this time constant will reduce rapidly due to exponential increase in sub-threshold leakage. These reasons make the stack effect based leakage reduction techniques attractive in nanoscale CMOS circuits.

REFERENCES

[1] S. Narendra, S. Borkar, V. De, D. Antoniadis, and A. Chandrakasan, "Scaling of stack effect and its application for leakage reduction," *Intl. Symp. Low Power Electronic and Design*, pp. 195-200, 2001.

[2] Y. Ye, S. Borkar, and V. De, "A Technique for Standby Leakage Reduction in High-Performance Circuits," *Symp. of VLSI Circuits*, pp. 40-41, 1998.

[3] D. Lee, W. Kwong, D. Blaauw, and D. Sylvester, "Simultaneous sub-threshold and Gate-Oxide Tunneling Leakage Current in Nanometer CMOS Design," *Intl. Symp. Low Power Electronic and Design*, pp. 287-292, 2003.

[4] A. Grove, http://www.intel.com/pressroom/archive/speeches/grove_ 20021210.pdf, *IEDM 2002 Keynote Luncheon Speech*.

[5] J.P. Halter and F. Najm, "A gate-level leakage power reduction method for ultra-low-power CMOS circuits," *Custom Integrated Circuits Conf.*, pp. 475-478, 1997.

[6] Z. Chen, M. Johnson, L. Wei, and K. Roy, "Estimation of Standby Leakage Power in CMOS Circuits Considering Accurate Modeling of Transistor Stacks," *Intl. Symp. Low Power Electronics and Design*, pp. 239-244, 1998.

[7] A. Chandrakasan, W. J. Bowhill, and F. Fox, *Design of High Performance Microprocessor Circuits*, IEEE Press, pp. 46-47, 2000.

[8] Z. Liu, C. Hu, J. Huang, T. Chan, M. Jeng, P. Ko, and Y. Cheng, "Threshold Voltage Model for Deep-Submicrometer MOSFET's," *IEEE Transactions on Electron Transistors*, vol. 40, no. 1, pp. 86-95, January 1993.

[9] Y. Taur, "CMOS Scaling beyond 0.1μm: how far can it go?" *Intl. Symp. on VLSI Technology, Systems, and Applications*, pp. 6-9, 1999.

[10] S. Thompson et. al., *Symp. VLSI Tech.*, pp. 69-70, 1999.

[11] S. Mutoh et. al, *IEEE JSSC*, pp. 847-854, Aug. 1995.

[12] T. Kuroda et. al, *IEEE JSSC*, pp. 1770-1779, Nov. 1996.

[13] L. Su, R. Schulz, J. Adkisson, K. Beyer, G. Biery, W. Cote, E. Crabbe, D. Edelstein, J. Ellis-Monaghan, E. Eld, D. Foster, R. Gehres, R. Goldblatt, N. Greco, C. Guenther, J. Heidenreich, J. Herman, D. Kiesling, L. Lin, S-H. Lo, McKenn, "A high-performance sub-0.25 μm CMOS technology with multiple thresholds and copper interconnects," *Intl. Symp. on VLSI Technology, Systems, and Applications*, pp. 18-19, 1998.

[14] D. T. Blaauw, A. Dharchoudhury, R. Panda, S. Sirichotiyakul, C. Oh, and T. Edwards "Emerging power management tools for processor design," *Intl. Symp. Low Power Electronics and Design*, pp. 143-148, 1998.

[15] S. Tyagi, M. Alavi, R. Bigwood, T. Bramblett, J. Bradenburg, W. Chen, B. Crew, M. Hussein, P. Jacob, C. Kenyon, C. Lo, B. Mcintyre, Z. Ma, P. Moon, P. Nguyen, L. Rumaner, R. Schweinfurth, S. Sivakumar, M. Stettler, S. Thompson, B. Tufts, J. Xu, S. Yang, and M. Bohr, "A 130 nm Generation Logic Technology Featuring 70 nm Transistors, Dual Vt Transistors and 6 layers of Cu Interconnects," *Intl. Elec. Transistors Meeting*, pp. 567-570, December 2000.

Chapter 3

POWER GATING AND DYNAMIC VOLTAGE SCALING

Benton Calhoun, James Kao[¶], and Anantha Chandrakasan
Massachusetts Institute of Technology, USA and [¶]Silicon Labs, Inc., USA

3.1 INTRODUCTION

This chapter examines *power gating* and *dynamic voltage scaling* as leakage reduction techniques. Both of these approaches use the power supply voltage, V_{dd}, as the primary knob for reducing leakage currents. Power gating refers to cutting off, or gating, a circuit from its power supply rails (V_{dd} and/or V_{ss}) during standby mode, and dynamic voltage scaling refers to changing the voltage supply (V_{dd} and/or V_{ss}) to achieve leakage savings.

Section 3.2 of this chapter describes power gating techniques. After describing generic power gating for leakage reduction, the section focuses on the Multi-Threshold CMOS (MTCMOS) power gating technique. We describe the significant issues related to MTCMOS design such as sleep transistor sizing and sequential logic design. Section 3.3 discusses the application of dynamic voltage scaling to the problem of leakage reduction. Section 3.3 describes the potential savings that standby voltage scaling can achieve and the methods for retaining state. It also examines using sub-threshold operation to reduce total energy.

3.2 POWER GATING

Power gating refers to using a MOSFET switch to gate, or cut off, a circuit from its power rail(s) during standby mode. The power gating switch typically is positioned as a header between the circuit and the power supply rail or as a footer between the circuit and the ground rail. During active operation, the power gating switch remains on, supplying the current that the circuit uses to operate. During standby mode, turning off the power gating switch reduces the current dissipated through the circuit. Since the switch gates the power when the circuit is in standby, it is also commonly called a sleep transistor.

For the most basic implementation of power gating, the sleep transistors are the same type of device as the transistors that implement the functional part of the circuit. Turning off the sleep transistor provides leakage reduction for two primary reasons. First, the width of the sleep transistor usually is less than the total width of the transistors being gated. The smaller width provides a linear reduction in the total current drawn from the supply during standby. Secondly, leakage currents diminish whenever stacks of transistors are off due to the source biasing effect. Many methods for reducing leakage leverage the source biasing effect [1], [2], [3].

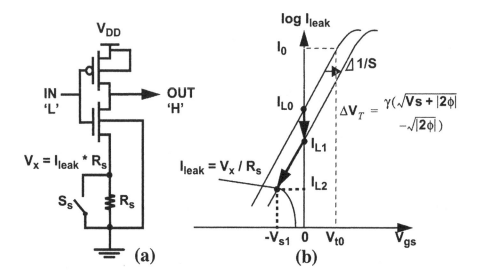

Figure 3-1. (a) Off-current reduction through the source biasing effect using switched source impedance. (b) Semilog plot of current shows the new operating point at the intersection of the resistor load line and the leakage current through the source biased NMOS.

In order to illustrate the source biasing effect, we will examine an early implementation of power gating called switched source impedance [1]. The switched source impedance approach uses an explicit resistor between the virtual ground rail of the circuit and the true ground node during standby as shown in Figure 3-1(a). A bypassing switch shorts out this resistor during active mode. As suggested in [1], the off resistance of the power gating switch serves as the switched source impedance in most power gating implementations. Source biasing refers to the change in voltage, V_x, at the source of the NMOS transistor(s) in the gated circuit. The sub-threshold leakage current plot in Figure 3-1(b) illustrates the reasons that source biasing decreases sub-threshold leakage. First, since V_x is greater than zero, the body effect raises the threshold voltage of the NMOS device(s) whose sources connect to that node. This increase in V_t shifts the log I_{leak} curve to the right such that the $V_{gs}=0$ leakage value moves from I_{L0} to I_{L1}. Secondly, the gate-source voltage of the same NMOS devices becomes negative, further reducing leakage currents to I_{L2}. The plot shows that the equilibrium voltage, V_x, is set where the leakage current through the gated circuit equals the current through the resistor, $I_{leak}=V_x/R$. When the off-resistance of the sleep transistor replaces the explicit resistance used with switched source impedance, V_x settles to equilibrium when the total leakage current into the drain of the sleep transistor matches the current through the circuit.

During active mode, the same effects cause a degradation of the circuit's speed. Even though the on-resistance of the power gating switch is much less than its off-resistance, it still creates a small positive voltage at the virtual node. Again, this voltage reduces the drive capability from $V_{gs}-V_t$ to $V_{gs}-V_t-V_x$ and increases the threshold voltage of the NMOS devices through body biasing. A more detailed discussion of the impact of the virtual rail voltage on speed follows in Section 3.2.2, which describes a common power gating approach.

3.2.1 Impact of power gating on gate leakage current

For older technologies, the total leakage current in an off transistor essentially equaled the sub-threshold current. However, technology scaling to nanometer dimensions has led to reduced oxide thicknesses to the point that gate current is on the same order as sub-threshold leakage. Since power gating transistors are typically quite large, their gate current threatens to offset some of the reduction in sub-threshold current that they provide. The total leakage current for a power-gated block becomes the sum of the sub-threshold leakage of the sleep transistor, the gate leakage of the sleep transistor, and the gate leakage of the input devices to the block. The gate current of the sleep transistor reduces its effective off-resistance in the example of Figure 3-1. This means that, for the same width, a sleep transistor with gate leakage causes a smaller V_x and thus reduced leakage savings in the source biasing scenario.

Scaling to nanometer technologies increases the impact of gate leakage on the MTCMOS technique. PMOS devices tend to exhibit lower gate leakage than NMOS, so using header sleep devices lowers the gate leakage relative to footer devices [14]. However, PMOS devices in nanometer technologies still have non-negligible gate current. One method to reduce the total leakage further is to apply an input vector to the block in standby to account for gate leakage [15]. For MTCMOS circuits using a footer sleep transistor, the internal nodes tend to float to V_{dd}. The opposite case occurs for PMOS header switches. In both cases, the gate-to-drain voltage of the sleep transistor approaches V_{dd}, resulting in the worst case gate leakage. Since the virtual ground of a block with an NMOS footer switch is close to V_{dd}, the drains and sources of most of the low Vt devices are also near the supply voltage. This means that the input stage to the block has maximum gate leakage for inputs of 0 and minimum gate leakage for inputs of V_{dd}. Applying the input vector 111...1 in this scenario minimizes gate leakage, providing savings of 40% to 50% relative to the average random input vector [15].

3.2.2 Multi-Threshold CMOS (MTCMOS)

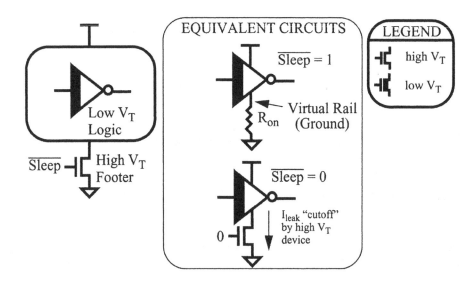

Figure 3-2. Basic MTCMOS structure with NMOS sleep device and equivalent circuits for active mode and for standby mode.

Multi-Threshold CMOS (MTCMOS) is a popular power gating approach that uses high V_t devices for power switches [4]. MTCMOS technology processing requires one or more additional implant steps to provide each additional threshold voltage. Figure 3-2 shows the basic MTCMOS structure, where a low V_t computation block uses high V_t switches for power gating. The figure shows a footer sleep transistor, but PMOS header switches or a combination of both header and footer switches are also used. When the high V_t transistors are turned on, the low V_t logic gates are connected to virtual ground and/or power. Their lower threshold voltage allows them to provide higher performance operation. However, by introducing an extra series device to the power supplies, MTCMOS circuits incur a performance penalty compared to CMOS circuits. When the circuit enters the sleep mode, the high V_t gating transistors turn off to reduce leakage currents.

The primary improvement of MTCMOS over generic power gating is the increased threshold voltage of the sleep transistor. Equation (1) shows the well-known equation for sub-threshold current in a MOSFET.

$$I_{subthreshold} = I_0 e^{\left(\frac{V_{gs} - V_t + \eta V_{DS}}{n V_{th}}\right)} \left(1 - e^{\frac{-V_{ds}}{V_{th}}}\right) \tag{1}$$

Clearly, the leakage current reduces exponentially for higher V_t. To first order, the leakage behavior of an MTCMOS circuit is characterized entirely by the threshold voltage of the sleep transistor. An additional small reduction in sub-threshold leakage occurs if the internal logic gates are configured to turn off the NMOS devices during the standby state to leverage the source biasing effect. The additional savings from source biasing are small compared to the leakage reduction from the higher threshold of the sleep device.

As with generic power gating, the voltage at the virtual node impacts the performance of the MTCMOS circuit. *Virtual rail bounce* refers to the changing voltage, V_x, at the virtual node(s) that occurs when active-mode current creates an IR drop across the on-resistance of the sleep transistor. For a footer sleep transistor, this rail bounce slows the output high-to-low transition by reducing V_{gs} of the discharging, low V_t NMOS from V_{dd} to $V_{dd} - V_x$ and by raising its V_t through the body effect since $V_{SB} = V_x$. In deep submicron devices, the lower V_{DS} will further reduce the discharge current because of Drain Induced Barrier Lowering (DIBL). Figure 3-3 shows generally why virtual rail bounce complicates MTCMOS circuit design [5]. Two different test vectors in the figure exercise the same critical path of the

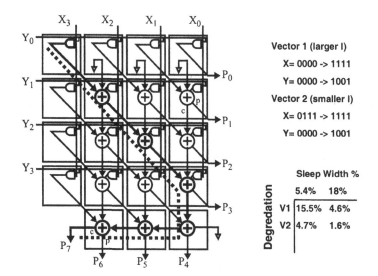

Figure 3-3. MTCMOS multipler showing dependency of delay on off-path transitions.

multiplier. The delay of the second vector is much larger due to the additional transitions off of the critical path. These off-path transitions result in extra current discharged through the sleep transistor, increasing the instantaneous virtual rail voltage. This simple example demonstrates that delay in MTCMOS circuits that share a sleep transistor depends on data transitions both on and off of the critical path.

For large chips with multiple power gated blocks, transitions to and from standby mode in one block can affect other blocks as well. Thus far, we have referred to the drain of the sleep transistor as the virtual node. For systems on a chip that have multiple power gated regions, the power rail at the source of the sleep transistor is also effectively a virtual power rail. This is due to the non-idealities in the bondpads and power distribution network. As a result of the self-inductance of bond wires and the parasitic inductance of the power grid, surges of current can cause the on-chip power rails to fluctuate [6], [7]. When a power-gated block is off, the internal nodes tend to discharge to ground (for a header sleep transistor). When the sleep transistor turns on, a large amount of current flows through the sleep transistor to charge up the internal nodes. This current surge through the bondpads and the power grid cause the inductive bounce we have described. The bouncing of the primary power rails of the chip affects all of its blocks. Any sensitive blocks with small noise margins can suffer spurious transitions or incorrect data as a result.

One technique for reducing bouncing for the on-chip power nodes is to increase the gate-to-source voltage of a sleep transistor in a step-wise manner. This gradual turn-on smoothes out the recharging of the nodes of the block leaving sleep. As a result, the current surge is reduced. This approach is shown to reduce the bounce of the on-chip supply by 60% to 80% for a test chip [6][7].

Despite some additional complexities, MTCMOS circuits are effective at solving the sub-threshold leakage problem during standby modes in nanometer technologies. Sleep mode operation is very straightforward, simply involving turning off the power switches, and will produce leakage reduction of up to several orders of magnitude, depending on the threshold voltages of the devices. On the other hand, the active mode circuit operation behaves theoretically just like an ordinary CMOS implementation, so existing architectures and designs can easily be ported to an MTCMOS implementation. One primary challenge for MTCMOS circuits and for power gating approaches in general is optimally sizing the sleep transistors. The next section explores issues with sleep transistor sizing on MTCMOS circuit performance, and techniques for optimally sizing sleep devices.

3.2.2.1 Sizing Sleep Devices

Correct sleep transistor sizing is a key parameter that affects the performance of MTCMOS circuits. If the sleep transistors are too large, then valuable silicon area would be wasted, switching energy overhead would be increased, and standby leakage reduction would be less than optimal. On the other hand, if the sleep transistors are too small, then the circuit speed becomes too slow because of the increased resistance to ground. Most sizing efforts attempt to use the minimum width possible to achieve a given performance requirement. Frequently, the performance constraint is a percentage of delay increase relative to the standard CMOS delay (i.e. – without sleep transistors). Let us define the optimum sleep size as the minimum sleep device width for which a circuit never exceeds a given delay.

One possible approach to estimate the transistor size is to sum the widths of internal low Vt transistors, but this can produce unnecessarily large estimates for transistor sizes. Ideally, one could simulate circuits for varying sleep transistor sizes with SPICE, but this can be very time consuming, especially if one tries to exhaustively test all possible input vectors for a complicated combinational circuit like an adder or multiplier. For large designs, exhaustive simulations quickly become prohibitive. Clearly, a better, more informative method of sizing the sleep transistor is necessary. The remainder of this section will address sleep transistor sizing issues and present several approaches from the literature.

Average Current Method

The Average Current Method (ACM) was proposed in [8] for sizing MTCMOS sleep transistors. This approach examines the average current of the active circuit and uses that value for sizing the sleep device. This technique uses only one sleep transistor. As we have shown, the delay of the circuit depends strongly on its discharge pattern when the virtual rail voltage changes significantly. It was observed, however, that this dependence was negligible for smaller values of incurred delay overhead [8] because the MTCMOS circuit looks like ordinary CMOS when the resistance of the sleep transistor is small. Thus, the ACM assumes that the delay penalty of inserting sleep devices is less than about 2%. This small delay penalty forces the maximum V_x to remain very small relative to V_{dd}, reducing the spread of delay coming from data dependencies. For this scenario, the ACM assumes that the current consumed is constant, and thus V_x is roughly constant. Application to a test chip showed a reduction in area from previous schemes that gave the same delay overhead [8]. Section 4.21.1 in Chapter 4 describes the ACM in greater detail and elaborates on its use in real applications.

This approach is limited by the requirement for such a small delay penalty. The dependence of delay on the data transitions makes the ACM ineffective for larger delays. This is because the delay can increase beyond the target value when the discharge current through the sleep device is larger than the average current. The ACM also has limited flexibility by assuming that only one sleep transistor should be used. The next section discusses some of the issues related to lumping sleep transistors into a few large devices or distributing many smaller sleep transistors locally.

Sizing for Global and Local Blocks

The decision to distribute many sleep transistors locally or to lump them into a few large transistors impacts the overhead, sizing, noise margins, and leakage reduction of MTCMOS circuits. A local sleep methodology inserts sleep devices at the local block or gate level. We define a local block as a section of the circuit that can be independently idle, such as a multiplier or a buffer chain. A global sleep methodology uses a single sleep transistor for a large circuit block containing multiple local blocks, such as an entire chip or a full data path.

For local sleep devices, exhaustive simulations can easily show that a gate or local block will always meet a given timing specification. The same guarantee becomes difficult to offer for large blocks without comprehensive simulation because the total circuit delay depends on the current drawn by the entire block. The 32-bit parity checker in Figure 3-4 provides a simple example to illustrate the effects of virtual rail bounce on local-style (right) and global-style (left) sleep devices [9]. The parity checker is really a local

block by our definition, but it behaves like a global block relative to the case
with sleep devices in each gate. The local-style sleep devices are sized so

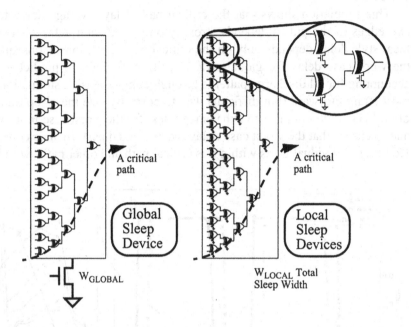

Figure 3-4. Example of a 32-bit parity checker using global (left) and local (right) sleep
methodologies. In practice, global blocks are often much bigger.

that the slowest transition of each XOR gate is slowed by less than 20%, so
the total delay penalty (average of all transitions) is close to 10%.

Figure 3-5 shows a simulation of the global-style parity circuit in which
two vectors are applied to the inputs. Vector 1 (0x00000001) activates only
the critical path, and Vector 2 exercises the critical path and switches some
of the off-path gates. Clearly, the off-path switching causes a larger ground
bounce. The large capacitance at the shared virtual rail filters the rail bounce
so that its peak value is lower than what might be expected. This tendency of
extra capacitance at shared virtual nodes to filter the rail bounce is one
incentive for sharing sleep transistors where possible. Although it decreases
the delay, this filtering can have a negative effect since the resulting slow tail
off of the virtual bounce can affect future transitions. For circuits with a
clock period close to the worst-case delay, the tail off can make delay
depend on the previous transition. Figure 3-5 also illustrates how global
sleep devices degrade noise margins in the close-up of the 'Out' signal. The
voltage bounce that appears in the 'Out' signal of the simulation occurs at
every node driven to '0', since the shared virtual ground node affects all
devices. Clearly, this erosion of noise margins could lead to errors in

sensitive circuits. Local sleep devices have better, more predictable noise margins [27] because their ground bounce depends on fewer discharging devices.

This simulation shows that the critical path delay (average delay for the two edges of the pulse) depends strongly on the off-path data, whereas the local-style delay depends only on the critical path transitions. Consequently, the worst-case delay for global-style circuits is difficult to predict without accounting for all discharge patterns through comprehensive simulation. The most straightforward (but difficult) way to correctly size the sleep transistor of an MTCMOS circuit is to exhaustively test for the worst case input vector and to ensure that the worst case delay meets a fixed performance constraint. However, individual gates within this critical path and other paths within the

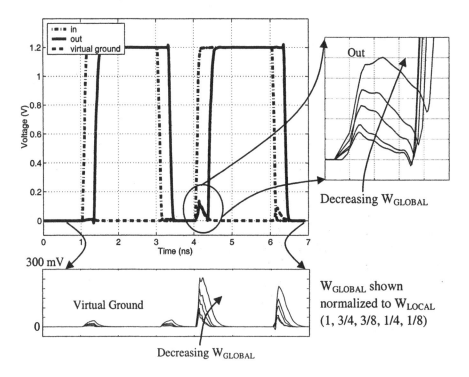

Figure 3-5. Simulation of the 32-bit parity checker showing delay along the critical path. The amount of virtual ground bounce for global-style sleep depends on the sleep device size (W_{GLOBAL}) and on off-path data (Vector 1, Vector 2) and decreases noise margins.

circuit can degrade in percentage more or less than this fixed criterion. A different way to satisfy a global performance criterion is to ensure that every individual gate meets a local performance constraint. This will ensure that any combination of gates in a path will also meet the performance requirements. Forcing every single gate to meet a nominal performance measure is a much more demanding constraint than simply meeting a global performance goal. However, in the context of MTCMOS circuits, it is much easier to implement this sizing strategy because one does not need to determine the worst case input vector pattern for the whole circuit. Instead, each individual gate can be assigned its own high V_t sleep transistor, whose size will be locally determined through exhaustive SPICE simulations.

This observation is leveraged to establish sizing methodologies that improve upon the ACM.

Hierarchical Transistor Sizing Methodology

The preceding discussion shows that it is easier to size a sleep transistor for a local gate than for a large block using exhaustive simulation. Suppose a circuit has optimum sleep devices distributed locally at the gate level. Summing these widths gives a sub-optimum global sleep device because, by superposition, the summed width is guaranteed to meet or exceed the performance of the local version even if all the gates discharge at once, which is unlikely. Reducing the size of the lumped sleep transistor requires additional thought.

Since not all gates in the circuit will switch at the same times, it is possible to merge sleep transistors from mutually exclusive gates together without increasing the circuit's delay [10]. A set of n gates with mutually exclusive discharge patterns and with equivalent sleep resistances $r_1, r_2,... r_n$ can share a single $r_{eff} = \min (r_1, r_2,... r_n)$. These mutually exclusive gates will discharge currents through the sleep transistor at different times so that the virtual ground bounce that each transitioning gate experiences will still be the same or smaller than before (from the extra capacitance at the virtual node). As a result, the delay of each gate sharing the common sleep transistor remains the same or decreases relative to the original circuit. An added benefit of replacing n sleep resistors with a single one is that the sub-threshold leakage current will decrease by the ratio of the new width to the total original width.

Although the MTCMOS transistor sizing algorithm has been presented at the gate level, in fact it can be applied at many hierarchical levels of a circuit. The algorithm simply operates on generic circuit blocks that are elements within a larger module, and each block is assumed to have a local high Vt sleep transistor that is used for gating the power supply rails. The algorithm is applied to the network by combining the sleep transistors for mutual exclusive blocks. Thus, the blocks that the algorithm operates on can represent individual gates, cells within an array (like an adder cell in a multiplier), or even a module within a chip (like an ALU). In all these cases, a gating sleep transistor can be shared among several different blocks if those blocks have activity patterns that do not overlap in time.

Using the hierarchical sizing methodology again, it is also possible to further apply this transistor merging technique on these existing modules into a larger system. However, by applying this nested algorithm at several levels of abstraction, we will tend to overestimate the minimum sleep transistor size required again, mainly because the larger granularity of the interactions between blocks makes it harder to find blocks with mutually exclusive discharge patterns. For example, applying the algorithm at the cell level within an array might give a larger estimate for the sleep transistor size than if the algorithm had been pushed down in the hierarchy and applied to the gates directly. However, utilizing a hierarchical approach to sizing the sleep transistors is very attractive because detailed circuit complexity can be abstracted away at the expense of accuracy, a tradeoff that is very often desirable.

Efficient Gate Clustering

A second clustering technique improves upon sharing sleep transistors with mutually exclusive discharge patterns by sharing sleep transistors for gates whose discharge patterns partially overlap. This approach, called efficient gate clustering, uses the desired delay penalty to determine the largest allowable virtual rail bounce, V_x. This information is used to size the sleep transistors for clusters of gates with exclusive and partially overlapping current discharge patterns [11][12][13]. The application of CAD algorithms combined with the intuition derived regarding mutually exclusive discharge currents makes a more efficient sizing technique.

The algorithm depends on a preprocessing step in which gates are clustered such that their combined current never exceeds the maximum current of any single gate in the cluster. To accomplish this, the propagation delays and expected discharge currents (through the sleep device) of the logic cells used in the design are fully characterized across different transitions, fanouts, etc. The expected discharge current of the gate is its peak discharge current multiplied by the probability that a discharging transition occurs. Using the transition statistics of the gate to lower its effective worst-case discharge current implicitly assumes that the different gates in a cluster will never discharge their worst case currents at the same time, thereby avoiding pessimistic sizing [13]. On the other hand, this approach makes it conceivable that the delay constraint will be exceeded if multiple gates in a cluster simultaneously discharge more than their expected value discharge currents.

The expected discharge pattern in time is modeled as a triangular pulse. The width of the pulse depends strongly on the fanout of the gate, and the peak of the pulse is the expected value of the discharge current. This pulse of discharge current can occur at different times if the inputs transition at different times. The triangular pulses representing discharge current for each input time are added together, resulting in a trapezoidal shaped pulse that represents the worst-case current discharged by the gate across all possible transition times. The preprocessing algorithm then groups gates together into clusters by examining the sum of the trapezoidal discharge current patterns. The algorithm prevents a new gate from joining a cluster when summing its discharge pattern to the pattern of the cluster creates a combined current larger than the peak current of any gate already in the cluster. The preprocessing algorithm results in clusters of gates with exclusive or partially overlapping discharge patterns whose combined current is less than the peak current of any single gate.

Once the preprocessing step concludes, the clusters of gates are packed more tightly into a smaller set of clusters whose peak current is set by the size of the sleep transistor. This repacking uses either a bin packing (BP) technique or a set partitioning (SP) approach. Both algorithms assign the clusters such that the peak current manageable by the sleep transistor is never exceeded by the sum of the currents of all of the gates. The BP and SP algorithms tend to produce more sleep transistors than the mutually exclusive approach, but the total sleep width is smaller. The BP approach performs very well for small, random circuits with unbalanced structures. It does not account for the placement of the gates on the chip. The SP technique makes up for this shortcoming, and its cost function accounts for the physical distance between gates in a cluster. This attention to reducing interconnect makes the SP algorithm attractive for nanometer technologies with increased interconnect capacitance. Using the BP and SP technique, the efficient gate clustering sizing approach produces smaller total sleep transistors than the average current method and the mutually exclusive clustering method for a variety of circuits [13].

3.2.2.2 Local Sleep Devices

Local sleep devices offer a significant opportunity for leakage savings by using local sleep regions to turn off small blocks when they are idle. Thus, the circuit can remain active at a global level, but unused blocks will draw reduced leakage current. In general, the local approach is preferred for ease of design or when local sleep regions can give active leakage savings that reduce total system leakage. Section 4.2.2.1 in Chapter 4 elaborates on the selective MT method, which uses standard cells that include local sleep transistors for implementing the critical paths in a circuit [17]. Also, the sizing approaches in the preceding sections have shown improvements by clustering local sleep transistors together. These methods begin with the approach of sizing the local sleep devices to meet the same performance constraint. Keeping the sleep devices separate at the local level allows CAD algorithms to relax the performance constraint for local gates while meeting the global constraint.

One new approach uses local sleep devices at the gate level without any clustering [18]. This method sizes the sleep transistors to meet a global delay requirement by taking up all of the available slack in the design. This means that the delay for many individual gates increases more than the global constraint, but the total worst-case delay meets the constraint. In addition to the delay constraint, an area constraint prevents the local sleep transistors from imposing too great of an area overhead. Despite this area constraint, the delay constraint remains the limiting factor for sizing the sleep devices. This distributed approach reports over 60% improvement in leakage reduction relative to clustering approaches with only 5% area overhead [18].

Another approach to distributed sleep transistors that minimizes the area overhead takes advantage of the area slack in a row of standard cells [19]. The rows of the post-layout design are compacted subject to a constraint on the routing congestion. The extra space generated by this step is collected at the edge of the rows and used to accommodate sleep transistors from a pre-characterized library. The sleep transistors are allotted to the gates in each row to minimize power consumption subject to the area and delay constraints. The fast reactivation time achieved by the local sleep devices contributes to minimizing the total delay of the approach and makes it ideal for active leakage reduction. This technique is reported to provide 80% leakage savings and 19% overall power savings with 2.5% area overhead [19].

The overhead of these local sleep devices requires some attention. Both local and global sleep transistors require routing new traces to every gate in the circuit: the sleep signal for the local sleep devices, and the virtual node for global devices. For a global sleep device, routing the virtual rail is equivalent to routing V_{dd} or ground, so wide wires are required to avoid resistive voltage drops. The interconnect resistance for distant blocks must be accounted for while sizing the sleep transistor in this scenario [13]. True ground might be routed with smaller traces, but it still needs to access most gates for substrate contacts. The sleep control signal can be minimum width for the local approach, so its routing overhead is always less than that of the global case. However, the sleep control circuitry for local sleep devices depends on the circuit while global designs only enter sleep when the entire circuit is idle. Although some local sleep designs will not require complicated control, random logic could demand complicated circuitry for determining when certain local blocks are in sleep. This overhead makes the local approach impractical for some circuits. However, most low power designs already employ clock gating methods for reducing active power [16]. Clock gating already requires signals that indicate when certain blocks are idle, so the presence of such signals could reduce the additional overhead for local sleep regions.

3.2.2.3 Power Domain Interfacing

The interfaces between different sleep regions require special attention. Power gating by nature creates floating nodes during the standby state. If a floating output voltage from a sleeping circuit drives the input of an active gate in an adjacent sleep region, a static current path from power to ground can result. Thoughtful partitioning of gates into clusters that will always be off together eliminates many of these hazards. The remaining interfaces between potentially sleeping regions with active regions can use circuits such as transmission gate multiplexers to actively drive the inputs to the active region and to prevent floating nodes from driving the gate of a transistor [9]. Other special circuits can provide a seamless boundary between power domains, such as the leakage feedback gate in Section 3.2.2.4.

An additional concern for power gating interfaces is sneak leakage paths. A sneak leakage path is any current path from V_{dd} to ground that continues to draw high current relative to a cutoff path during sleep mode. Sneak leakage paths can occur whenever an MTCMOS output node is connected electrically to another node with low impedance to a power rail. This electrical connection most often occurs through low Vt transmission gates, but sneak paths may also occur through structures such as clock-gated inverters and tri-state buffers. Formally, a sneak leakage path is a current path that flows from V_{dd} to ground through a set of "on" devices, **A**, and through a set of "off" devices, **B**. Set **B** contains only low Vt devices, while Set **A** contains low and/or high Vt devices [9]. Since the off devices are low Vt, the sneak leakage current is roughly one or more orders of magnitude higher than other currents in the circuit, and floating nodes can drive them much higher. The higher leakage current lets a few sneak paths erode the savings achieved by cutting off many other paths, so sneak leakage paths are a prohibitive problem that must be prevented to make local sleep devices feasible.

It may seem unlikely that sneak paths would exist in a carefully designed circuit, but the term *sneak leakage* implies that they can be quite subtle. The basic structure of MTCMOS circuits suggests that sneak leakage paths can occur only where the sleep device(s) can be bypassed, at the interface between MTCMOS and CMOS type circuits [24]. A conservative approach to ensure that Set **B** always contains a high V_t device could use both polarities of sleep device for each MTCMOS gate without sharing any sleep devices. This approach incurs a large area penalty over the optimum sleep device size by using many unnecessary sleep devices.

Proper design techniques can permit designers to approach the optimum sleep device size without allowing sneak leakage even in circuits using transmission gates for speed. Four design rules allow placement of sleep devices while preventing sneak leakage [9]. The first rule states that a shared output (through low V_t transmission gate) between a high V_t gate and an MTCMOS gate can be prevented by using both polarity sleep devices. This type of sneak leakage path might occur at the input of a flip-flop that has a high V_t feedback inverter. It also appears when MTCMOS logic on the critical path interfaces with high V_t logic off of the critical path through a transmission gate multiplexer. The second rule requires MTCMOS gates with shared outputs to have the same polarity sleep device(s). This type of sneak path can appear when a design is optimized for minimum leakage using both polarity sleep devices. For example, if a known input is applied to a circuit during sleep mode, then the outputs at every node are determined prior to asserting sleep. Local sleep devices can be selectively placed to force stacks of off devices at each logic stage for extra leakage reduction. Such a design approach could create the leakage path in the second rule. The third rule states that a gate with a shared sleep device must have the same polarity sleep device(s) as the other gate. As previously mentioned, local sleep devices for gates with mutually exclusive discharge patterns can be shared to reduce area. If sleep device widths are optimized in this way and then rule 1 violations are fixed, this type of sneak leakage path can occur. The fourth rule prevents a special sneak leakage path that might occur for a circuit that complies with the first three rules. In this case, a sleep device is shared between two MTCMOS gates that in turn share outputs with high V_t gates. This path could arise if a designer tries to reduce area by sharing sleep devices for the input buffers to several flip-flops. Sleep devices should not be shared when they connect multiple high V_t outputs.

3.2.2.4 Sequential Logic

Although MTCMOS circuit techniques are effective for controlling leakage currents in combinational logic, a drawback is that it can cause internal nodes to float, so data stored in sequential elements is lost. As a result, the literature contains several possible MTCMOS latch designs that can reduce leakage currents yet maintain state during the standby modes.

This section describes several useful designs from literature that highlight the key approaches to sequential MTCMOS design.

Basic Master-Slave Flip-Flop

An MTCMOS flip-flop that can retain memory during the standby state is shown below in Figure 3-6 [24]. This implementation is a straightforward extension of a standard master slave flip flop, where leakage paths are carefully eliminated.

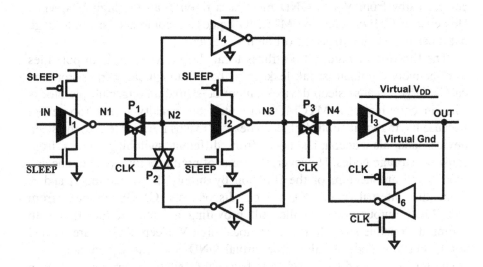

Figure 3-6. MTCMOS flip-flop using parallel high Vt structures.

The basic latch structure used in this flip flop uses low V_t devices throughout the critical path while high V_t devices are used in circuits parallel to the critical path that simply hold state. One drawback to this approach is the extra capacitance added to the critical path. By making passgate P_2 high V_t, the performance of the master latch actually improves because during the transparent state, I_1 would not have to fight against I_5 through an off low V_t passgate. However, the drawback is that when CLK goes high, I_5 and P_2 need to be strong enough to correctly hold the state at node N_2 because N_2 and N_1 might be driven to opposite rails. Other than this sizing precaution, the active operation of this flip flop is straightforward. During the standby state, all leakage current paths are also eliminated to minimize power dissipation.

One of the problems with sequential circuits that utilize feedback and parallel devices is that sneak leakage paths may exist. The flip-flop of Figure

3-6 is a good example of how distributed high V_t sleep transistors and dual polarity sleep devices are needed to eliminate sneak leakage currents during the standby condition. Sneak leakage paths can arise in MTCMOS circuits whenever the output of an MTCMOS gate is electrically connected to the output of a CMOS gate. In fact, the interfacing between MTCMOS type circuits and CMOS type circuits is what gives rise to potential leakage paths. For example, if a datapath block is implemented with only MTCMOS gates, then a single high V_t switch (either PMOS or NMOS) is sufficient to eliminate sub-threshold leakage currents during the standby state because all current paths from V_{dd} to GND must pass through an off high V_t device. However, if CMOS gates and MTCMOS gates are combined, sneak leakage paths can arise that bypass the off high V_t devices.

The flip-flop of Figure 3-6 utilizes local sleep devices of both polarities to effectively eliminate sneak leakage paths. Although sleep transistor area can be large because sleep devices cannot be shared among multiple blocks (like in combinational MTCMOS circuits), the penalty is not too severe because having local control of sleep devices makes it easier to size the sleep devices and also decouples noise from different switching blocks from sensitive storage nodes. However, if area is of premium importance, one can modify the architecture of the flip flop by simply disconnecting I_4 and I_5 from the internal node N_3, which disconnects CMOS outputs from MTCMOS output nodes, while still providing a latch recirculation path during the opaque state. In this case, local high V_t sleep devices are needed for I_1, but a shared virtual V_{dd} or virtual GND line with a common sleep transistor can be used for I_2 and I_3. In fact, only a single polarity shared sleep device is needed to eliminate leakage currents. In a large register for example, sharing a common NMOS sleep transistor among several flip flop and logic blocks can result in large area savings, at the expense of more complicated sleep transistor sizing methodologies.

Balloon Latch

Another MTCMOS sequential circuit that holds state during the idle mode is the "balloon" circuit described in [22], [23]. Instead of using parallel high Vt inverters to maintain recirculation paths during the sleep state, this approach uses a completely autonomous balloon circuit that is used to explicitly write in stored data during the standby state, and can be read out when returning to the active mode. These balloon circuits can be made minimum sized because they simply hold data and do not need to be fast. The other benefit is that all MTCMOS gates can share common virtual V_{dd} and virtual ground lines since MTCMOS gates are completely decoupled from the high V_t CMOS balloon elements. The schematic of an MTCMOS balloon circuit is shown in Figure 3-7.

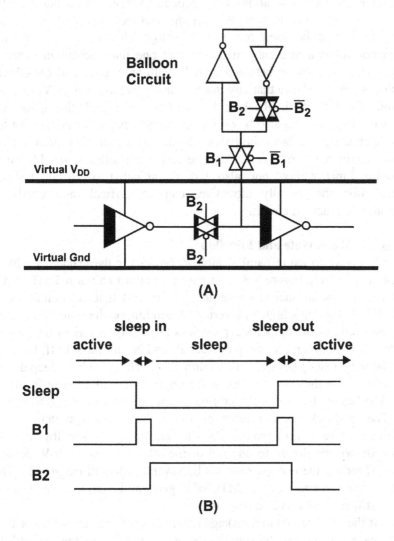

Figure 3-7. MTCMOS balloon circuit schematic (A) and control signals (B).

The balloon circuit operates in four distinct phases. During the active mode, the balloon circuit is disconnected from the internal MTCMOS logic through a high V_t passgate. During the sleep-in stage, the data on the internal MTCMOS node is stored into the balloon circuit. During the sleep state, the balloon circuit is again disconnected from the MTCMOS logic, and the data is re-circulated in the high V_t balloon circuit. Finally, during the sleep-out state, the MTCMOS logic path is broken and the stored data is written into the MTCMOS node. Returning back to the active mode completes the cycle.

Although the basic operation is theoretically simple, the balloon circuit approach still requires a complex timing methodology and redundant circuitry that must be used for each flip-flop. This extra circuitry and complex control is a necessary trade-off that one must accept in order to provide a clean separation between high V_t balloon circuits and MTCMOS logic blocks, and to ensure that any interactions between the high V_t and low V_t circuits are decoupled. Routing the extra control signals throughout the chip to each flip flop can also be costly in a large design. A variation of the balloon latch design replaces the balloon latch with a scan-able latch in [25]. The scan chain registers provide data retentive capabilities in addition to scan chain functionality, but also has complicated overhead. Balloon circuits, while theoretically attractive, can be difficult and costly to implement in a practical circuit.

Leakage Feedback Gate and Flip-flop

An alternative to using parallel high V_t devices is the leakage feedback gate [24] as shown in Figure 3-8. This gate is similar to a normal MTCMOS gate, but it has additional sleep devices (P_2 or N_2) that are conditionally turned off by a feedback high V_t inverter. Depending on the state of the latest output, only one of the helper sleep devices (P_2 or N_2) is turned on. During the sleep state, both high V_t sleep devices P_1 and N_1 are turned off, but only one of the helper sleep devices will be turned off. The on helper sleep device continues to drive the output signal to the appropriate rail, as seen in Figure 3-8(a). The figure shows that the original output of the gate is a zero during sleep. The feedback inverter turns on the high V_t footer switch, N_2, to maintain an active path to ground. As a result, during the standby state the output node will be driven to one rail or the other, but the high V_t devices that are off reduce the leakage currents by several orders of magnitude. This provides a mechanism where an MTCMOS gate can be put in a low leakage state, yet still actively drive its outputs.

Even if the input signal to a leakage feedback gate does transition or float after the gate is placed in the standby mode, the output voltage will still be held to the same logic value through a leakage path. Figure 3-8(b) illustrates this effect. The original output of the gate was zero, as seen in Figure 3-8(a). If the input to the gate transitions from V_{dd} to 0, then the active path to ground through N_2 is cutoff. Now there are two leakage currents that affect the output node voltage. One leakage path through the low V_t PMOS (on) and P_1, P_2 (off) is set by high V_t devices. The second leakage current through the off high V_t devices N1 and N2 is orders of magnitude less powerful due to the higher V_t. The differences in V_t allow the pull-up leakage current to dominate and thus to hold the output voltage to the same rail as before. Because the functionality of this leakage feedback selectively enables either the virtual power or virtual ground lines depending on the last data present, leakage feedback gates must utilize local high V_t sleep devices. However, utilizing local sleep devices may already be desirable for sensitive circuits like flip-flop and latches that should be decoupled from the switching activities of neighboring gates.

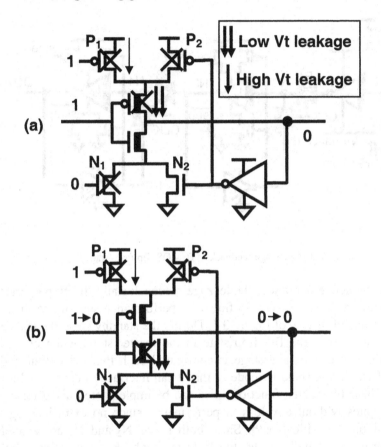

Figure 3-8. (a) Leakage feedback gate during sleep. (b) When the input changes, the larger low V_t leakage current pins the output at its original value.

The leakage feedback gate can be used directly as an interface circuit between MTCMOS and CMOS logic blocks. For example the last stage of any MTCMOS block can be implemented as a leakage feedback gate such that during the standby mode the output is still driven. As a result, this stage can safely drive a standard CMOS gate without creating short circuit currents due to floating inputs.

Another use of the leakage feedback gate is to modify the static MTCMOS flip-flop to eliminate the need for the parallel inverter to recirculate data. The leakage feedback structure can be used instead, which does not slow down the critical path because no extra capacitances are introduced on internal nodes as seen in Figure 3-9. The addition of the helper sleep devices only add load to the outputs of I_4 and I_5 which are not part of the critical path, so the speed of the MTCMOS flip flop is not compromised.

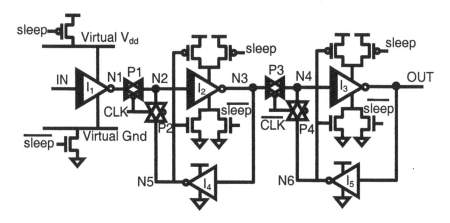

Figure 3-9. Leakage feedback MTCMOS flip flop (static).

During the active operation, the leakage feedback static flip-flop operates like an ordinary master slave flip flop with performance superior to that of the master-slave flip-flop of Figure 3-6. During the standby state, the leakage feedback gate stores the flip flop state in the master stage while clock is high. Furthermore the slave stage is configured such that the output node does not float during standby mode so that it can interface directly to CMOS devices. These two added functionalities can be implemented using leakage feedback gates without any loss in performance since no extra loading is introduced on the critical path. Since both nodes N_2 and N_4 are actively driven during the standby state, the leakage feedback gates actively hold their outputs as well, and thus functions exactly like the previous flip-flop using high V_t re-circulation paths.

Leakage feedback gates also can implement dynamic flip-flops that retain state during the standby mode yet still have very fast circuit performance during the active state, as shown in Figure 3-10. During the active switching mode, the leakage feedback dynamic flip flop operates like a conventional one, where the state of the master and slave latches are simply stored on the dynamic nodes at the inputs of I_1 and I_2. Because leakage currents will be large when using low V_t devices, the clock period must be made fast enough such that the node voltages can be stable over the clock periods of interest. By utilizing leakage feedback gates in a dynamic flip flop however, it will be possible to retain data during the standby state when CLK is high. This enables a timing methodology where standby gating can be performed so that a block can be stalled in time, and yet can be woken up again to complete the computation. Clock gating alone would not be sufficient because during the active state the flip flop's dynamic nodes would simply leak away, and data would be corrupted. The leakage feedback gate provides an added functionality to architectures with dynamic flip flops not available before. Conventional dynamic flip flops are incapable of maintaining data during the standby modes, so architectures that provide for block shut down periods must explicitly provide peripheral circuitry to retain state.

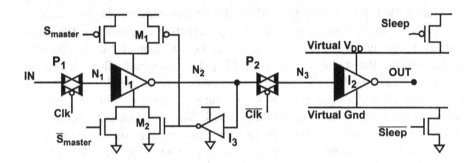

Figure 3-10. Leakage feedback dynamic flip-flop.

When the dynamic flip-flop is placed in a sleep condition however, the leakage feedback mechanism retains the data even if the internal dynamic nodes change. However, because the input voltage to the leakage feedback gate can float, the timing requirement coming out of the sleep state requires that the phase of the master latch sleep signals lag the main sleep signals by one half cycle. This is because immediately turning on the master latch high Vt sleep transistors when transitioning from the sleep state to the active mode might cause the data to flip state. As a result, the master slave can exit the sleep condition only after the clock goes low and the slave stage latches the stored data.

3.2.3 Modifications on Power Gating

The increased emphasis on low power design has encouraged the development of numerous modifications to the basic power gating and MTCMOS design methodologies. These modifications typically improve upon the standard approaches for a specific type of application requirement. This section briefly describes several of the major proposed modifications to power gating.

One simple way to improve the conductance of the sleep transistors in the active mode is to overdrive the gate voltage (i.e. above V_{dd} for the NMOS device) [26]. The extra gate drive in active mode either allows a smaller sleep transistor to support the same performance or provides a higher performance for the same leakage reduction capability. One limitation to this approach is the prevalence of gate current in nanometer technologies. As described in [26], the gate oxide of the sleep transistor must be thick enough to prevent excessive gate leakage upon application of the higher gate voltage. As long as the gate leakage problem is solved, this approach offers significant improvements over standard power gating. Also, since most chips have higher supply voltages already (i.e. – for IO), the overhead of generating the higher supply for NMOS sleep transistors may be decreased.

A similar method improves the leakage reduction capability of the sleep transistors by under-driving the gate (i.e. below ground for the NMOS device) during the standby mode [20], [27]. This approach allows a low Vt transistor to act as the sleep device, improving circuit speed during active mode and allowing operation at lower supply voltages without reduced performance relative to standard MTCMOS. The negative gate voltage in sleep mode prevents the leakage current from getting out of hand. One limitation to this approach may be GIDL current if the drain-to-source voltage is too large. As with the active gate overdrive, this approach may require positive or negative charge pumps, so the overhead deserves careful attention.

Zigzag Super Cut-off CMOS combines alternating header and footer sleep transistors with overdriving the sleep transistor gate [21]. This approach alternates between gating to virtual ground and virtual V_{dd} for gates that are tied in series. This configuration improves the wake-up time of the sleeping blocks by 8 times over the standard gate overdrive approach [21]. Section 4.3 of Chapter 4 provides a detailed look at these advanced circuit approaches for power gating.

3.3 DYNAMIC VOLTAGE SCALING

Dynamic voltage scaling typically refers to changing the voltage supply and operating frequency of a circuit to minimize power by reducing slack time. While this technique reduces dynamic power in a scalable way, it generally does not address leakage power. This section describes the application of dynamic voltage scaling to reducing leakage power in a circuit.

3.3.1 Standby leakage reduction

While power gating is a popular approach to reducing standby leakage power, dynamic voltage scaling offers another option [28]. Dynamic voltage scaling for standby leakage reduction refers to lowering the supply voltage by either reducing V_{dd} or by raising V_{ss}.

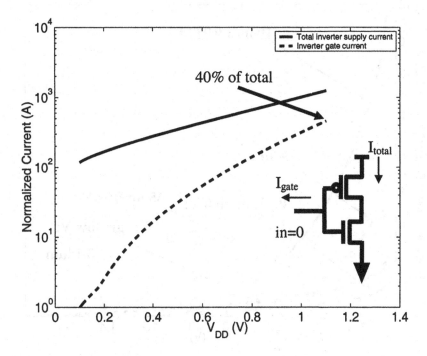

Figure 3-11. Contribution of gate leakage to total leakage versus V_{dd} in 90nm CMOS.

Lowering the supply voltage when the circuit is idle reduces leakage power in several ways. Clearly, the lower voltage itself provides a linear reduction in power. The current also decreases. Specifically, Equation (1) shows the expression for sub-threshold leakage.

This equation indicates that lowering V_{dd} will produce a corresponding exponential reduction in sub-threshold current resulting from Drain-Induced Barrier Lowering (DIBL) factor, η. At extremely low V_{DS} values ($\sim kT/q$), the parenthetical term produces a more pronounced roll-off in leakage current. Lowering V_{dd} thus saves standby power by decreasing both sub-threshold current and V_{dd}. Lowering V_{dd} also reduces gate leakage even faster than sub-threshold leakage [28]. The plot in Figure 3-11 shows how the contribution of gate leakage to total leakage in a 90nm process rolls off quickly as V_{dd} decreases. The rapid decrease in tunneling gate current at lower gate voltages results from the smaller number of available tunneling electrons. Moreover, gate induced drain leakage (GIDL) is not an issue at low values of V_{DS}, and simulations of a 90nm process indicate that diode leakage remains small relative to the sub-threshold current in the sub-threshold region.

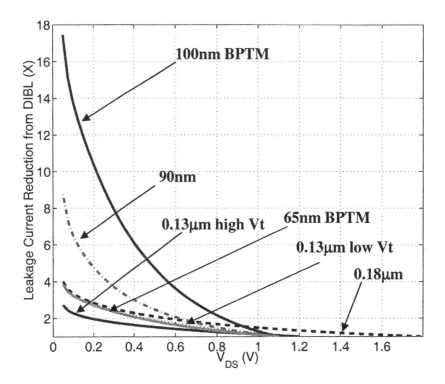

Figure 3-12. Decrease in sub-threshold current due to DIBL in 3 commercial technologies and 2 Berkeley Predictive Technology Model (BPTM) technologies [29][30].

Since the dominant component of leakage at lowered V_{dd} is sub-threshold current, the amount of total power savings from voltage scaling depends on the DIBL factor. Figure 3-12 plots normalized I_{DS} versus V_{DS} for $V_{gs}=0$ in three commercial technologies and two Berkeley Predictive Technology Models [29][30]. The plot shows that the DIBL effect reduces current by roughly 2X to 4X for each technology when V_{dd} scales from its nominal value to about 300 mV. Thus, the total theoretical power savings are on the order of 8X to 16X for scaling down to the 300 mV range if the nominal V_{dd} is 1.2V. Dramatic additional savings arise from the roll off in sub-threshold current at even lower supply voltages.

Figure 3-12 gives some insight into the scalability of dynamic voltage scaling for leakage power reduction in nanometer technologies. The DIBL effect tends to become more severe with process scaling to shorter gate lengths. For this reason, the savings achievable by this technique will increase with technology scaling.

We have shown qualitatively that lowering V_{dd} will provide power savings, but it is helpful to quantify those savings. Presumably, the standby power supply for a circuit can decrease to zero, but the circuit will lose all of its state. The optimal point for power savings using this technique is the lowest voltage for which the circuit retains state. In theory, the combinational logic in a circuit does not need to hold state. If two power supplies are readily available, the voltage supply to combinational logic can fall all the way to zero while sequential elements use a different power supply. The sequential supply may be decreased to save power, but it must remain above some minimum point to hold its state. Separating the power supplies increases power savings dramatically because all of the combinational logic draws zero power in standby while the sequential logic uses reduced power resulting from a lower V_{dd}. Separating the power supplies to the combinational and sequential logic is often impractical, so we will assume that the entire circuit under test shares the same power supply.

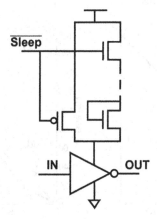

Figure 3-13. Schematic for self-controllable-voltage level standby voltage scaling.

One approach to standby voltage scaling uses diode stacks together with power gating MOSFETs to pinch in the rail voltages during standby mode, as shown in Figure 3-13 [31]. The quantity and sizes of the devices used in the diode stack determine the reduced V_{dd} value and the increased V_{ss} value during standby mode. In order to preserve state, the total voltage drop across the circuits (V_{ds}) is decreased by about 40%. This implementation of standby voltage scaling is conservative, and a greater reduction of V_{dd} is preferable for reducing standby power. At the same time, careful treatment of a circuit's storage elements is necessary for state preservation.

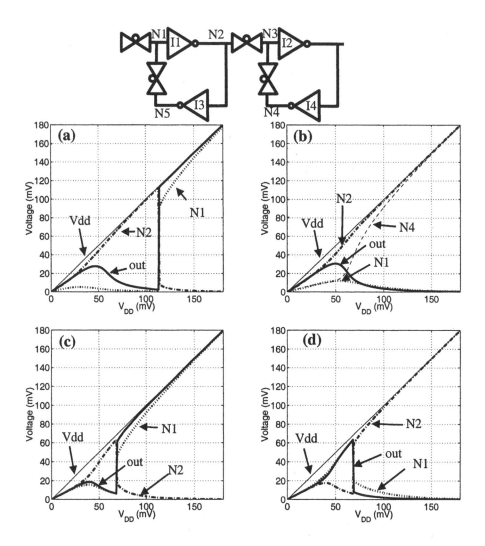

Figure 3-14. (a) DC sweeps of master-slave flip-flop showing data retention for an unbalanced flip-flop holding a 1, (b) unbalanced flip-flop holding a 0, (c) balanced flip-flop holding a 1, and (d) balanced flip-flop holding a 0.

A primary concern for standby voltage scaling is retaining the state stored in the flip-flops and memory elements. Figure 3-14 shows a DC sweep for a master-slave flip-flop in 0.18μm technology [34]. The sizing of the inverters in the flip-flop affects the minimum voltage for which the circuit holds state. Figure 3-14(a) shows an unbalanced flip-flop losing the '1' state at 100mV, and (b) shows the same flip-flop retaining a '0' all the way to 0 volts, even with the worst-case input. Balancing the sizes in the flip-flop can minimize the voltage for holding both states, as shown in the Figure 3-14(c) and (d). Measurements of leakage feedback flip-flops on a 0.13 μm test chip confirm state retention to 95 mV [33] at room temperature. Measurements of master-slave flip-flops on a 90 nm test chip show state retention to 110 mV at room temperature. Analysis of an SRAM showed a slightly higher result. The data retention voltage was measured to be 250 mV at room temperature for an SRAM in [32].

Clearly, the minimum voltage for these flip-flops will fluctuate with process, voltage, and temperature (PVT) variations. One approach to minimizing the standby voltage without losing state across PVT variations is using canary flip-flops [33][34]. These are flip-flops sized to lose their state at a higher voltage than the flip-flops used to store data. Simulations have confirmed that they consistently fail at higher voltages than the critical flip-flops across process and temperature variation [34]. Using these flip-flops permits standby voltage scaling close to the point of failure without actually losing data.

3.3.2 Subthreshold Operation for Total Energy Minimization

Clearly, leakage power is consumed during both active and standby mode. A majority of leakage management strategies focus on reducing standby leakage power. For some applications, however, energy consumption is a larger concern than performance. For example, medical devices, microsensor nodes, wake-up circuitry on DSPs, and rarely used local blocks all have relaxed performance requirements and would ideally use as little energy as possible. When minimizing energy is the primary system requirement, the sub-threshold region gives the minimum energy solution [35][36] for most circuits. Subthreshold circuits use a supply voltage, V_{dd}, that is less than the threshold voltage, V_t, of the transistors. In this regime, sub-threshold leakage currents charge and discharge load capacitances, limiting performance but giving significant energy savings over nominal V_{dd} operation. Figure 3-15 gives an example of sub-threshold operation for a 0.18μm ring oscillator. The left-hand plot shows the on and off currents for an NMOS device versus the supply voltage. For nominal V_{dd}, the ratio of I_{on} to I_{off} is quite large. Once V_{dd} drops into the sub-threshold region, the current becomes exponential with voltage, and the I_{on}/I_{off} ratio reduces quickly. The right-hand plot shows FO4 ring oscillators retaining 10% to 90% output swing at V_{dd}=80 mV. The performance is exponentially lower than at nominal V_{dd} due to the lower on-current.

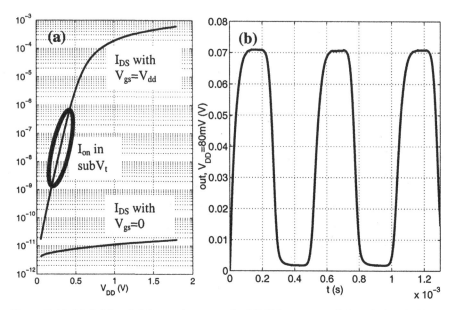

Figure 3-15. (a) Subthreshold operation example: NMOS current and (b) sub-threshold ring oscillator operation at 80 mV for a 0.18 μm CMOS technology.

The advantage of sub-threshold operation is the dramatic reduction of energy. It is well-known that active energy scales as the switched capacitance times the square of the supply voltage. These quadratic savings in energy continue until leakage energy becomes dominant. The increasing leakage energy occurs because of the increased delay in the sub-threshold region that allows longer integration time for the leakage current during each cycle. Models for total energy in sub-threshold coincide with these results [37][38]. An analytical solution for the optimum supply voltage to minimize energy is derived in [37]. That model is based on a characteristic inverter in the process of interest. The delay of the characteristic inverter is given in the following equation:

$$t_d = \frac{KC_g V_{dd}}{I_{o,g} e^{\frac{V_{gs}-V_{t.g}}{nV_{th}}}} \qquad (2)$$

The K term is a fitting constant, C_g is the switched capacitance of the inverter, and $I_{o,g}$ and $V_{t.g}$ are analogous to the MOSFET parameters of the same name. The frequency of operation for a generic circuit is just $f=1/(t_d L_{DP})$, where L_{DP} is the logic depth in terms of characteristic inverter delays. Combining these equations with fitting constants for matching to an arbitrary circuit gives the equation for total energy per cycle [37]:

$$E_{Total} = C_{eff}V_{dd}^2 + W_{eff}L_{DP}KC_gV_{dd}^2 e^{-\frac{V_{dd}}{nV_{th}}}$$

$$= V_{dd}^2\left(C_{eff} + W_{eff}KC_gL_{DP}e^{-\frac{V_{dd}}{nV_{th}}}\right) \tag{3}$$

This total energy equation extends the expressions for current and delay of an inverter to arbitrary larger circuits, sacrificing accuracy for simplicity since the fitted parameters cannot account for all of the details of every circuit. Thus, C_{eff} is the average total switched capacitance of the entire circuit, including the average activity factor over all of its nodes. Likewise, W_{eff} estimates the average total width that contributes to leakage current, normalized to the characteristic inverter. Treating this parameter as a constant ignores the state dependence of leakage, but gives a good average estimate of the chip-wide leakage. Averaging the circuit leakage current for simulations over many states improves the total leakage estimate.

A parallel 8-bit, 8-tap FIR filter provides an example for observing the minimum energy point in sub-threshold operation. Simulations of the filter use netlists extracted from synthesized layout in a 0.18 μm process. The synthesis flow incorporates a standard cell library modified to enable operation down to 100 mV at the typical corner. Calibrating the model requires three parameters: C_{eff}, L_{DP}, and W_{eff}. C_{eff} is calculated from a simulation of the circuit performing typical operations using $C_{eff} = I_{avg} / (f*V_{dd})$. L_{DP} is the delay of the critical path divided by the delay of the characteristic inverter, and W_{eff} is the average leakage current of the circuit divided by the average leakage current of the characteristic inverter. Figure 3-16 shows the simulated minimum energy point for the FIR filter. The leakage energy remains negligible until the sub-threshold region, at which point it begins to increase exponentially. The rise in leakage energy occurs because of the increased delay, since the leakage current actually decreases from DIBL. For the FIR, the optimum supply voltage is 250 mV at a frequency of about 20 kHz [37].

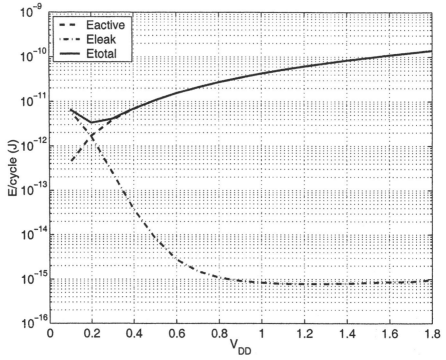

Figure 3-16. Minimum energy point for 8x8 FIR filter.

The minimum energy point does not remain constant for a given circuit. Instead, it is influenced by environmental changes such as temperature and by operating changes such as duty cycle. Any relative increase in the leakage energy per cycle pushes the optimum V_{dd} higher, and the frequency at the optimum point also increases. This corresponds to any decrease in active energy or increase in leakage energy. Likewise, any decrease in leakage energy or increase in active energy lowers the optimum V_{dd}.

Designing sub-threshold circuits has several limitations. First, sub-threshold circuits suffer from high sensitivity to process variations because of the exponential dependence of current on V_t. Fine-grained threshold voltage control can counteract the effects of process variation using adaptive back biasing or leakage controlled feedback circuits. Secondly, the minimum energy point is sensitive to variables such as activity factor, temperature, duty cycle, and sizing. This sensitivity makes careful analysis of a circuit's usage important for selecting the optimum V_{dd}, and it suggests the benefits of closed loop control of V_{dd}. The low V_{dd} also makes the circuits more prone to soft errors, although redundancy techniques can compensate for critical circuits.

3.3.3 Comparison with Power Gating

Power gating gives improvements in standby power by reducing leakage current by from 10X to several orders of magnitude. Power gating offers less reduction of leakage for solutions that minimize the delay overhead incurred by the sleep transistors. Dynamic voltage scaling during standby mode offers on the order of 50% to 80X power savings. The lower savings are for less voltage scaling, and the larger savings are from aggressive voltage scaling in nanometer technologies with large DIBL effect. The energy savings achievable by sub-threshold operation are roughly one order of magnitude, since they are determined by the ratio of the active energy at the nominal voltage to the total energy at the optimum voltage.

One advantage to standby voltage scaling for leakage reduction is that the technique is easily applied in cooperation with power gating. For example, the voltage supply to a power gated circuit can be scaled to a lower value. In this scenario, the reductions in standby power offered by the two approaches are multiplied to give an estimate of the total savings. Clearly, a strategy combining power gating with standby voltage scaling can reduce standby leakage by many orders of magnitude.

3.4 SUMMARY

This chapter has described the basic application of power gating and dynamic voltage scaling for reducing leakage power in nanoscale CMOS. Power gating techniques have become very common in literature and in practice, and MTCMOS implementations in particular have demonstrated significant improvements in standby power consumption. Dynamic voltage scaling also offers improvements in leakage power. Lowering V_{dd} during standby mode saves leakage power, and sub-threshold operation during active mode can minimize total energy per operation.

REFERENCES

[1]   M. Horiguchi, T. Sakata, and K. Itoh, Switched-source-impedance CMOS circuit for low standby sub-threshold current giga-scale LSIs, *IEEE Journal of Solid-State Circuits*, Vol. 28, No. 11, pp. 1131 – 1135, November 1993.

[2]   T. Kawahara, M. Horiguchi, Y. Kawajiri, G. Kitsukawa, and T. Kure, Subthreshold current reduction for decoded-driver by self-reverse biasing, *IEEE Journal of Solid-State Circuits*, Vol. 28, No. 11, pp. 1136-1144, November 1993.

[3]   Y. Ye, S. Borkar, and V. De, A new technique for standby leakage reduction in high-performance circuits, *1998 Symposium on VLSI Circuits*, pp. 40-41, June 1998.

[4]   S. Mutoh, T. Douseki, Y. Matsuya, T. Aoki, and J. Yamada, 1V high-speed digital circuit technology with 0.5μm multi-threshold CMOS, *IEEE International ASIC Conference and Exhibit*, pp. 186-189, September 1993.

[5]   J. Kao, S. Narendra, and A. Chandrakasan, Subthreshold leakage modeling and reduction techniques, *International Conference on Computer Aided Design*, pp. 141-148, November 2002.

[6]   S. Kim, S. V. Kosonocky, D. R. Knebel, K. Stawiasz, D. Heidel, and M. Immediato, Minimizing inductive noise in system-on-a-chip with multiple power gating structures, *European Solid-State Circuits Conference*, pp. 635 – 638, September 2003.

[7]   S. Kim, S. V. Kosonocky, and D. R. Knebel, Understanding and minimizing ground bounce during mode transition of power gating structures, *IEEE International Symposium on Low Power Electronics and Design*, pp. 22-25, August 2003.

[8]   S. Mutoh, S. Shigematsu, Y. Gotoh, and S. Konaka, Design method of MTCMOS power switch for low-voltage high-speed LSIs, *Asia and South Pacific Design Automation Conference*, pp. 113-116, January 1999.

[9]   B.H. Calhoun, F.A. Honore, and A.P. Chandrakasan, A leakage reduction methodology for distributed MTCMOS, *IEEE Journal of Solid-State Circuits*, Vol. 39, No. 5, pp. 818-826, May 2004.

[10] J. Kao, S. Narendra, and A. Chandrakasan, MTCMOS hierarchical sizing based on mutual exclusive discharge patterns, *Design Automation Conference*, pp. 495-500, June 1998.

[11] M. Anis, M. Mahmoud, and M. Elmasry, Efficient gate clustering for MTCMOS circuits, *IEEE International ASIC/SOC Conference*, pp. 34-38, September 2001.

[12] M. Anis, S. Areibi, M. Mahmoud, and M. Elmasry, Dynamic and leakage power reduction in MTCMOS circuits using an automated efficient gate clustering technique, *Design Automation Conference*, pp. 480 – 485, June 2002.

[13] M. Anis, S. Areibi, and M. Elmasry, Design and optimization of multithreshold CMOS (MTCMOS) circuits, *IEEE Transactions on Computer-Aided Design of Integrated Circuits and Systems*, Vol. 22, No. 10, pp. 1324 – 1342, October 2003.

[14] F. Hamzaoglu and M. Stan, Circuit-level techniques to control gate leakage for sub-100nm CMOS, *IEEE International Symposium on Low Power Electronics and Design*, pp. 60-63, August 2002.

[15] R. Rao, J. Burns, and R. Brown, Circuit techniques for gate and sub-threshold leakage minimization in future CMOS technologies, *European Solid-State Circuits Conference*, pp. 313-316, September 2003.

[16] C. Bittlestone, A. Hill, V. Singhal, and N.V. Arvind, Architecting ASIC libraries and flows in nanometer era, *Design Automation Conference*, pp. 776-781, June 2003.

[17] K. Usami, N. Kawabe, M. Koizumi, K. Seta, and T. Furusawa, Automated selective multi-threshold design for ultra-low standby applications, *IEEE International Symposium on Low Power Electronics and Design*, pp. 202-206, August 2002.

[18] V. Khandelwal and A. Srivastava, Leakage control through fine-grained placement and sizing of sleep transistors, *IEEE/ACM International Conference on Computer Aided Design*, November 2004.

[19] P. Babighian, L. Benini, A. Macii, and E. Macii, Post-layout leakage power minimization based on distributed sleep transistor insertion, *IEEE International Symposium on Low Power Electronics and Design*, pp. 138-143, August 2004.

[20] H. Kawaguchi, K. Nose, and T. Sakurai, A CMOS scheme for 0.5V supply voltage with pico-ampere standby current, *International Solid-State Circuits Conference*, pp. 192-193, February 1998.

[21] K. S. Min, H. Kawaguchi, and T. Sakurai, Zigzag super cut-off CMOS (ZSCCMOS) block activation with self-adaptive voltage level controller: an alternative to clock-gating scheme in leakage dominant era, *International Solid-State Circuits Conference*, pp. 400-401, February 2003.

[22] S. Shigematsu, S. Mutoh, Y. Matsuya, and J. Yamada, A 1-V high-speed MTCMOS circuit scheme for power-down applications, *Symposium on VLSI Circuits*, pp. 125-126, 1995.

[23] S. Shigematsu, S. Mutoh, Y. Matsuya, Y. Tanabe, and J. Yamada, A 1-V high-speed MTCMOS circuit scheme for power-down application circuits, *IEEE Journal of Solid-State Circuits*, Vol. 32, No. 6, June 1997.

[24] J. Kao and A. Chandrakasan, MTCMOS sequential circuits, *European Solid-State Circuits Conference*, pp. 332-339, 2001.

[25] V. Zyuban and S. Kosonocky, Low power integrated scan-retention mechanism, *IEEE International Symposium on Low Power Electronics and Design*, pp. 98-102, August 2002.

[26] T. Inukai, M. Takamiya, K. Nose, H. Kawaguchi, T. Hiramoto, and T. Sakurai, Boosted gate MOS (BGMOS): device/circuit cooperation scheme to achieve leakage-free giga-scale integration, *IEEE Custom Integrated Circuits Conference*, pp. 409-412, 2000.

[27] M. Stan, Low threshold CMOS circuits with low standby current, *IEEE International Symposium on Low Power Electronics and Design*, pp. 97-99, August 1998.

[28] R. Krishnamurthy, et al, High-performance and low-power challenges for sub-70nm microprocessor circuits, *IEEE Custom Integrated Circuits Conference*, 2002.

[29] http://www-device.eecs.berkeley.edu/~ptm

[30] Y. Cao, T. Sato, D. Sylvester, M. Orshansky, and C. Hu, New paradigm of predictive MOSFET and interconnect modeling for early circuit design, *IEEE Custom Integrated Circuits Conference*, pp. 201-204, Jun. 2000.

[31] T. Enomoto, Y. Oka, H. Shikano, and T. Harada, A self-controllable-voltage-level (SVL) circuit for low-power, high-speed CMOS circuits, *European Solid-State Circuits Conference*, pp. 411-414, September 2002.

[32] H. Qin, Y. Cao, D. Markovic, A. Vladimirescu, and J. Rabaey, SRAM leakage suppression by minimizing standby supply voltage, *IEEE International Symposium on Quality Electronic Design*, pp. 55-60, 2004.

[33] B. Calhoun and A. Chandrakasan, Standby voltage scaling for reduced power, *IEEE Custom Integrated Circuits Conference*, pp. 639-642, September 2003.

[34] B. Calhoun and A. Chandrakasan, Standby power reduction using dynamic voltage scaling and canary flip-flop structures, *IEEE Journal of Solid-State Circuits*, Vol. 39, No. 9, September 2004.

[35] A. Wang, A. Chandrakasan, and S. Kosonocky, Optimal supply and threshold scaling for sub-threshold CMOS circuits, *Symposium on VLSI*, pp. 5-9, April 2002.

[36] J. Burr and A. Peterson, Ultra low power CMOS technology, *3rd NASA Symposium on VLSI Design*, pp. 4.2.1-4.2.13, 1991.

[37] B. H. Calhoun and A. Chandrakasan, Characterizing and modeling minimum energy operation for sub-threshold circuits, *IEEE International Symposium on Low Power Electronics and Design*, pp. 90-95, August 2004.

[38] B. Zhai, D. Blaauw, D. Sylvester, and K. Flautner, Theoretical and practical limits of dynamic voltage scaling, *Design Automation Conference*, pp. 868-873, June 2004.

Chapter 4

METHODOLOGIES FOR POWER GATING

Kimiyoshi Usami[§] and Takayasu Sakurai[¶]

[§]*Shibaura Institute of Technology, Japan and [¶]University of Tokyo, Japan*

Power gating is a technique to reduce leakage current by electrically disconnecting the circuit from the power and/or the ground using a power switch. While the principle and the structure are very simple, there are issues to be solved when applying it to real designs. Verification and validation methods on the circuit performance are complex when the power switch is shared among logic gates. This is because the critical path delay is affected by discharge patterns of logic gates (i.e. input vectors). Methodologies to give practical solutions to these issues have been developed and applied to real designs in industry. This chapter describes power gating methodologies employed in real designs and presents future directions of power gating techniques.

4.1 INTRODUCTION

Power gating is one of the most effective techniques to reduce leakage energy dissipation that increases exponentially with device scaling. Since the Multi-Threshold voltage CMOS (MTCMOS) technique [1] was proposed, extensive researches have been conducted from both circuit and methodology perspectives.

In the original MTCMOS, the power switch is shared among low-Vth logic gates, as depicted in Figure 4-1(a). The basic principle of MTCMOS is presented in Chapter 3.2. Determining the proper transistor width for the power switch is a very important task because it gives a big impact to performance, area and power. However, this task is complicated and difficult because the gate delay is affected by discharge patterns of other gates that share the power switch [2]. In other words, critical path delay of the circuit depends on input vectors. In a real product design, verification to guarantee the timing is essential. However, exhaustive timing verification for all possible input vectors is impractical for large designs. How to guarantee the performance is a significant issue when the power switch is shared?

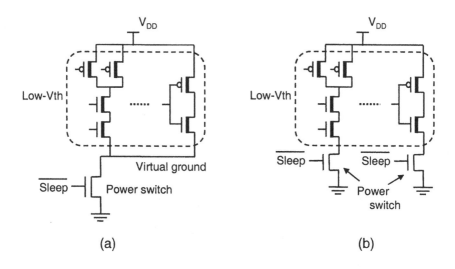

Figure 4-1. Shared and unshared power-switch schemes. (a) Shared. (b) Unshared.

The most straightforward way to avoid this problem is to "unshare" the power switch, as shown in Figure 4-1(b). Since the power switch is not shared among the logic gates, the gate delay is not affected by discharge patterns of other gates. This makes it easy to determine the proper power-switch size. By characterizing each gate delay taking into account the power switch, delay analysis on the critical paths and timing verification can be performed as in the normal design. On the other hand, the drawback of this scheme is area penalty. If every gate is equipped with the power switch, the total area overhead due to the power switch will be unacceptable. How to minimize the area penalty is a key issue in the unshared scheme.

This chapter describes methodologies for the shared and unshared power-switch schemes that are applied to real designs in industry. For the shared scheme, the Average Current Method (ACM) [3] is employed in the NTT's communication LSI [3] and the Samsung's PDA processor [4]. Since the basic idea on the ACM is presented in Chapter 3.2, this chapter describes the methodology on how to determine the proper power-switch size by using the ACM more in detail. For the unshared scheme, the Selective MTCMOS (or the Selective MT) technique [5] is used in the Toshiba's DSP core for W-CDMA cell phones [5] and the Qualcomm's cellular baseband chip [9].

The methodology to minimize the area penalty by using the Selective MT technique is described.

In addition to those existing techniques, future directions of the power gating are discussed as well in this chapter.

4.2 POWER GATING METHODOLOGIES FOR REAL DESIGNS

4.2.1 Methodologies to share the power switches

In real MTCMOS designs to share the power switch, the Average Current Method (ACM) is used as a practical methodology to determine the proper power switch size. This chapter describes an analytical way to determine the power switch size and presents design examples employing the ACM.

4.2.1.1 Average Current Method (ACM)

The ACM is a methodology to determine the power switch width based on the average current of logic circuits that share the power switch [3]. Let us look at how the gate delay in MTCMOS is affected by discharge patterns of other gates. Figure 4-2(a) shows a circuit to evaluate this influence. The circuit consists of a gate chain to measure the delay and a power consuming block. The gate chain and the power consuming block share the power switch. When the power consuming block discharges, the voltage of the virtual ground VSSV raises to higher than 0V due to the discharging current and the ON resistance of the power switch. The increase of the virtual ground voltage by ΔV reduces the effective supply voltage of the gate chain to V_{DD}-ΔV. This in turn degrades the performance and increases the gate delay. Simulations have been conducted to examine the delay penalty by changing operating patterns while keeping the average current of the power consuming block constant. Three different operating patterns are generated by changing the configuration in the power consuming block. The power consuming block is modeled as a circuit consisting of N lines of M-stage gate chains, as shown in Figure 4-2(b). The number of gate stages M and the number of lines N are changed while keeping M*N constant. Figure 4-2(c) shows examples of the operating patterns. From the simulation results shown in Figure 4-2(d), it is found that the delay penalty does not strongly depend on the operating patterns if a power switch with sufficient size is used and hence the speed penalty is constrained to be sufficiently low (e.g. less than 1.02 in Figure 4-2(d)).

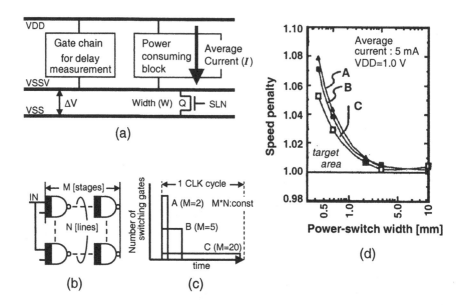

Figure 4-2. Average Current Method. (a) Simulation circuit. (b) Power consuming block.
(c) Operating pattern. (d) Simulation results on speed penalty.

Based on this observation, the ACM has been proposed. In the ACM it is
assumed that the current dissipation is constant in the circuit to which
MTCMOS is applied, and hence ΔV is constant. This assumption is
appropriate because of the fact that the delay penalty is almost independent
of the operating patterns when targeting at sufficiently low delay penalty.
This makes it easy to analyze the delay penalty since the voltage drop at the
power switch and performance degradation can be treated as static
phenomena.

An analytical way to derive the required minimum power-switch size
from the given delay penalty is as follows. The gate delay at the supply
voltage V_{DD} is expressed as

$$\tau[V_{DD}] \propto \frac{C \cdot V_{DD}}{\beta(V_{DD} - Vtl)^{\alpha}} \tag{4.1}$$

where β is the drivability factor, C is the output load capacitance, α is the
saturation index, and Vtl is the low Vth.

As previously demonstrated, the effective supply voltage of the power
gating circuit is V_{DD}-ΔV. It is assumed that gate delay is equal to that of the

low-Vt CMOS circuits with the V_{DD}-ΔV power supply. The MTCMOS speed penalty *MSP* expresses the ratio of the delay times at V_{DD}-ΔV and V_{DD} as

$$MSP = \frac{\tau[V_{DD} - \Delta V]}{\tau[V_{DD}]} \cong \left[\frac{V_{DD} - Vtl}{V_{DD} - Vtl - \Delta V}\right]^{\alpha-1} \qquad (4.2)$$

Now ΔV is obtained by

$$\Delta V = RI = \frac{R'}{W}I \qquad (4.3)$$

where R and I are the power switch resistance and the average current flowing the power switch, respectively. R' is the normalized power switch resistance and the actual R is determined by dividing R' with the power switch width W. By rearranging Eq. (4.2) by substituting Eq. (4.3) into it, the *MSP* can be rewritten as

$$MSP = \frac{1}{\left(1 - \frac{I}{W}\cdot\frac{R'}{V_{DD} - Vtl}\right)^{\alpha-1}} \qquad (4.4)$$

The power switch width W is finally expressed as

$$W = \frac{1}{1 - \sqrt[1-\alpha]{MSP}}\cdot\left(\frac{R'}{V_{DD} - Vtl}\right)\cdot I \qquad (4.5)$$

By knowing the average current I, the power switch size W is determined.

Equation (4.5) is convenient in roughly estimating the power switch size. However, it contains several approximations. In order to determine the power switch size more accurately, the following procedure is used instead. First, two basic data are prepared: (i) the delay penalty dependency on ΔV, and (ii) the I-V characteristics of the power switch near $V_{ds}=0$. Examples of (i) and (ii) are shown in Figure 4-3 (a) and (b), respectively. The resistance of the power switch can be derived from the slope of the graph in Figure 4-3 (b). In this example, R is approximately 3 Ω with the gate width of 1mm, leading to R' of 3 Ω-mm. The required power switch size for the given delay penalty is derived in the following way. If a 2% delay penalty is allowed, ΔV is approximately 0.02 V from Figure 4-3 (a). When the average current I is 5 mA, the required power-switch size can now be determined using Eq. (4.3) as

$$W = \frac{R'I}{\Delta V} = \frac{3[\Omega mm] \times 5[mA]}{\Delta V} = \frac{15[mVmm]}{\Delta V}$$ (4.6)

For ΔV=0.02 V, the power switch size W is calculated as 0.75 mm. For other delay-penalty values specified, the power switch size W can be determined as well in the same manner.

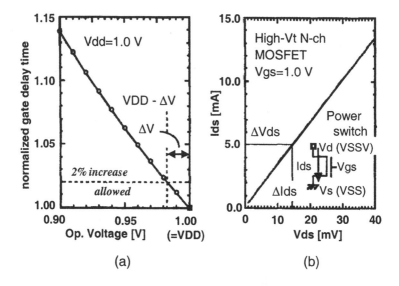

Figure 4-3. Basic data for the ACM. (a) gate-delay dependence on the voltage.
(b) I-V characteristics of the power switch.

4.2.1.2 Design examples of Average Current Method

- **NTT's communication LSI [3]**

    The ACM has been applied to a 290K gates communication LSI designed using 0.25 μm MTCMOS/SIMOX process. It is reported that the power switch width determined using the ACM is 190 mm under the condition of the average current 210mA and the delay penalty 1.02. Observation in the measurement shows that the delay penalty is 2% for the 190 mm power switch. Compared to the conventional method requiring the 1000 mm width, the ACM reduces leakage power from 6 μW to 1 μW.

- **Samsung's PDA processor [4]**

The ACM has been used at the power-gating design of a 32-bit microprocessor for personal digital assistants (PDA's). The paper [4] reports that a design flow has been built for MTCMOS using commercially available tools. After synthesizing the RTL code, power consumption is estimated to know the average current. The aggregate width of power switches is determined using the ACM. Based on this width and floor-planning information, power switch cells are inserted into the optimal locations while taking into account the ground bounce. This methodology for MTCMOS has been evaluated at a 16-bit DSP and a PDA processor for 0.18μm CMOS technology. It is reported that 342 power switch cells with the size of 5μm have been inserted at a 958 x 957 μm² module. The transistor level simulation for the DSP indicates that the ground bounce is 9mV on average and varies up to 49 mV at the 1.8 V supply voltage, resulting in degrading performance by 2%. In the fabricated PDA processor, the measured leakage power in the sleep mode is approximately 2μW. This number is 6000 times smaller than that of the non-MTCMOS implementation.

4.2.2 Methodologies to unshare the power switches

The scheme to unshare the power switch has a great advantage that power-switch sizing is simple and timing verification can be performed as in the non-MTCMOS design. Meanwhile, the drawback is area overhead. Since providing power switches for all the gates causes intolerable area penalty, an approach to selectively apply MTCMOS in the circuit has been proposed, leading to a Selective Multi-Threshold (MT) technique [5].

4.2.2.1 Selective MT technique

In the Selective MT technique, multi-threshold (MT) cells to which power gating is introduced are employed only in the critical paths while high-Vth cells are used in the non-critical paths [5], as shown in Figure 4-4. The MT cell is composed of low-Vth logic transistors and a high-Vth power switch. Basic structure of the MT cell is depicted in Figure 4-5.

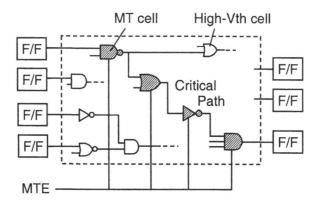

Figure 4-4. Selective MT circuit.

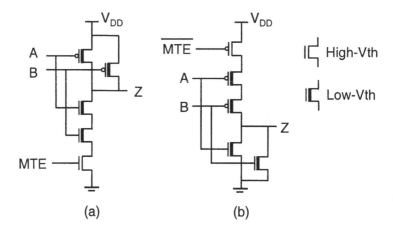

Figure 4-5. Basic structure of MT cell. (a) 2-input NAND cell. (b) 2-input NOR cell.

The MT cell has a control input "MTE" (MT Enable) to switch the operation between active mode and standby mode. In the active mode, MTE is set to '1', resulting in performing fast logic operation with low-Vth transistors in the MT cell. In the standby mode, MTE is set to '0'. The high-Vth power switch is turned off, resulting in cutting off the sub-threshold leakage path from VDD to ground. In addition to this structure, a circuit to avoid output-floating at MTE='0' is incorporated into the MT cell, as described later.

In the Selective MT technique, the MT cells are employed to speed up the critical paths. This technique has the following advantages. First of all, it does not have a virtual ground line appeared in the shared power switch scheme. Hence, the gate delay is not affected by the discharging pattern of other gates. The size of the power switch can be independently determined to be optimal within an MT cell. Second, mode switching between active and standby can be performed at one clock cycle. Also, power consumption at the mode switching is lees than the shared power switch scheme. Since power switches exist only in MT cells on critical paths, the total transistor width of power switches is much smaller than that of the shared scheme. Because of this, static power in the active mode is reduced as well in the Selective MT. Lastly, the data stored in flip-flops and latches are maintained even at standby mode because high-Vth cells are used for them. It does not require special flip-flop circuits (e.g. a balloon circuit [6]) which are used in the shared power switch scheme.

The cell library and the design flow to support the Selective MT technique are described below.

- **Cell Library**

The MT cells are essential components in the Selective MT. However, they do not exist in the conventional ASIC library. It is reported in [5] that instead of developing MT cells corresponding to the entire set of the ASIC library they chose MT cells to develop under the following policies. First, the cells with small drive strength are excluded. This is because the MT cells are employed in the critical paths, and hence cells with small drive strength are not likely to be used. Next, flip-flops and latches are excluded. High fan-in gates such as 8-input gates are also excluded because these can be realized by a combination of 2-input or 3-input gates. Finally, complex gates that are expected to contribute to speed up the critical paths are added. Based on these policies, 56 MT-cells including inverters, buffers, NANDs, NORs and several complex gates with variations of drive strength have been developed.

As depicted in Figure 4-4, MT cells and high-Vth cells are cascaded in a Selective MT circuit. In the structure shown in Figure 4-5, the output of the MT cell may become floating at MTE='0'. This may cause direct current path at high-Vth cells locating at the fan-out of the MT cells. To avoid this problem, a holding circuit to maintain the output voltage at MTE='0' is added to the MT cell. As an MT cell with an output-holder, "latch-type" and "bypass-type" circuits shown in Figure 4-6 are proposed in [5]. In the latch and bypass portions high-Vth transistors with the minimum transistor width are employed. At the design of the MT cells, either of a latch-type or a bypass-type circuit is chosen for each cell considering which type would give smaller area. For example, a bypass-type is chosen for an inverter and 2-input NAND and NOR gates because it gives smaller area than the latch-type for those gates. For 2-input AND and OR gates the latch-type circuit is chosen. The power-switch width has been determined as 3X the size of low-Vth transistors. By using this size, the performance of the MT cell with 0.55V and 0.35V Vth's becomes almost equal to that of the original cell with 0.45V Vth in 0.18μm CMOS technology.

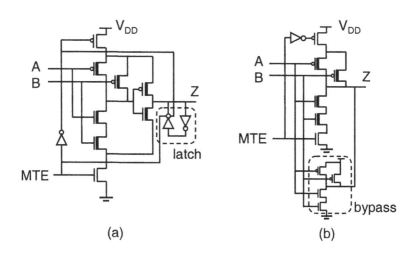

Figure 4-6. MT cell with output-hold circuit. (a) latch-type. (b) bypass-type.

- **Design Flow**

The Selective MT circuit shown in Figure 4-4 is not synthesized with commercially available logic synthesis tools even though both MT cells and high-Vth cells are given as the target library. The main reason is that even if logic function is the same at both an MT cell and a high-Vth cell the number of pins is different from each other. In other words, a high-Vth cell for NAND2 has two input pins (A and B) and an output pin (Z), while an MT cell for NAND2 has three input pins (A, B and MTE) and an output pin (Z). Hence, conventional tools are not capable of synthesizing a circuit in such a way that MT cells are mapped to the critical paths while high-Vth cells to non-critical paths.

In order to solve this problem, the following design flow has been proposed [5]. First, logic synthesis from the RTL code is performed using a conventional tool with high-Vth cells. Next, a Selective MT circuit depicted in Figure 4-4 is generated from the output of logic synthesis. In [7] it is described that the authors developed a tool identifying critical paths in a high-Vth circuit, and replacing high-Vth cells with MT cells so the entire circuit can meet the timing constraints. Cell replacement is performed from the primary outputs toward primary inputs using a backward traversal algorithm [8]. By using this algorithm, high-Vth cells closer to the primary outputs are more likely to be replaced with MT cells. This means MT cells are located in late stages in the critical paths. This makes the arrival time constraints on the MTE signal less critical. Since generation and propagation of the MTE signal take a certain delay time, the cell replacement in a backward fashion makes sense.

In the layout of a Selective MT circuit, routing the signal MTE is a key process. Because the signal MTE has a large number of fan-outs, connecting them only with metal wire and driving with a single buffer may cause problems such as electro-migration. To cope with this problem, a buffer-tree structure is automatically generated for the signal MTE in a clock-tree-synthesis (CTS) fashion. This enables to build an optimal buffer-tree structure taking into account the location of MT cells. In practice, since close consideration of skew matching is not required for the MTE signal, only the tree construction and the buffer placement are performed.

4.2.2.2 Design examples of Selective MT technique

As real design examples using the Selective MT technique, the Toshiba's DSP core for W-CDMA cell phones and the Qualcomm's cellular baseband chip have been reported in the literature.

- **Toshiba's DSP core for W-CDMA cell phones [5]**
The DSP core used in a baseband chip for cell phones requires very low standby current because it is directly related to the standby time. In the standby mode, cell phones intermittently page the base station to exchange information on the location, wireless channels, synchronization, etc. This operation is repeated at a certain period, as shown in Figure 4-7. It should be noted that the time spent for paging is only 3% in the period and remaining 97% is a sleep time [7]. Hence, reducing the standby leakage current in the sleep time is a key. In applications without the intermittent operation, power shutdown at standby mode to reduce leakage might be possible. However, in cell phones, power shutdown is accompanied by a significant overhead at every mode change from active to standby and vice versa. This is because saving the memory data before shutdown and restoring it after the wake-up are required. This overhead is not acceptable for cell phones in timing and power.

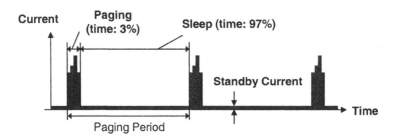

Figure 4-7. Intermittent operation in cell phones.

The paper [5] reports that the use of high-Vth (0.55V) in the entire circuit was needed to meet the standby leakage requirement for the DSP core in the 0.18μm CMOS technology and the supply voltage of 1.5V. However, the timing constraints for 100MHz cannot be met if only high-Vth cells are employed. In order to speed up the critical paths in the active mode and suppress the leakage in the standby mode, the Selective MT technique has been applied to the module with the highest timing-criticality. As the result of employing the Selective MT technique, 12% of high-Vth cells have been replaced with MT counterparts in the module containing 34K cells. Figure 4-8(a) shows a photograph of the test chip of the DSP core. The Selective MT technique has been introduced to a part of random logic located in the center of the chip. The fabricated chip operated at 100MHz. Area overhead is 10% in the part to which the technique was applied.

Figure 4-8(b) shows the results on the measured leakage current. In order to evaluate effectiveness of the Selective MT, leakage current has been measured at MTE='0' and at MTE='1' for the same chip. At MTE='0', leakage current at the standby mode is observed because the power switches in the MT cells are turned off. In contrast, at MTE='1' the leakage current is observed in the state that the power switches are turned on. It is found that the leakage current at the standby mode is reduced to 1/2-1/3 by using the Selective MT technique.

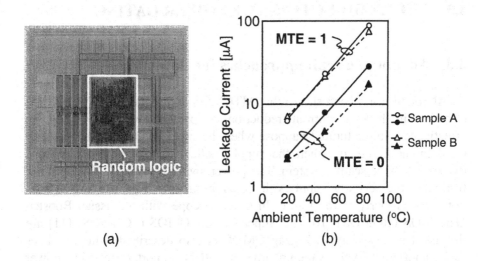

Figure 4-8. (a) Test chip for Selective MT. (b) Results on leakage measurement.

- **Qualcomm's cellular baseband chip [9]**

    Another real example of the Selective MT technique is the Qualcomm's 3G CDMA2000 1X cellular baseband chip. It contains GSM, AMPS, GPS, Bluetooth and multimedia capabilities, and is implemented in a 130nm dual-Vt CMOS process. The chip integrates a 32b ARM926 RISC processor, two DSP's and dedicated hardware accelerators. The RISC processor is required to operate at 180MHz at the ambient temperature of 95°C for the slow process corner. The required clock frequencies for the application DSP and the modem DSP are 95MHz and 85MHz, respectively. In order to reduce leakage power while meeting the performance goals, footswitch standard cells (i.e. cells with power switches) have been employed in the critical paths.

    In the layout, both non-footswitch and footswitch cells can be placed in the same row. This guarantees that the selective insertion of footswitch cells via automatic synthesis generates area-efficient layouts. The flip-flops use high-Vth without footswitches to obtain both low leakage and state retention. It is reported that the performance loss of each footswitch cell when compared to its low-Vth counterpart is 10%. Although the average footswitch cell is 1.25X the area of its non-footswitch counterpart, there is only a 4% chip area impact since the footswitch cells are only used in the critical paths and 66% of the chip consists of embedded RAMs and ROMs. It is reported that the use of the footswitches improves the standby time by 3X-4X.

4.3 FUTURE DIRECTIONS OF POWER GATING

4.3.1 Advanced circuit approaches for power gating

Although it has been shown that MTCMOS and selective MTCMOS are effective in leakage current reduction at present, will they keep its effectiveness in the future? Suppose when the supply voltage is decreased to 0.5V in the future or when the supply voltage is hopped to 0.5V in a dynamic voltage scaling system. The power switch with 0.5V V_{TH} does not turn on in the very low supply voltage environments. Thus MTCMOS does not work properly in the future. In order to cope with this issue, Boosted Gate MOS (BGMOS) [10] and Super Cut-off CMOS (SCCMOS) [11] are discussed in this chapter. Zigzag CMOS is also described, which realizes less-than-a-clock-cycle wake-up time. It will be important in using power gating as a substitute of clock gating in the future when the clock gating loses its effectiveness as the leakage is getting dominant in power consumption.

4.3.1.1 Power Gating in Lower Voltage (BGMOS and SCCMOS)

The circuit diagram of Boosted Gate MOS (BGMOS) is shown in Figure 4-9. In BGMOS, high drivability of a power switch is maintained by boosting the gate voltage of the power switch to the higher voltage than V_{DD}, say 2V. The high gate voltage can be either generated on chip or introduced from the external environments. The high voltage, however, may cause damage to gate oxide. To prevent the damage, the gate oxide should be thicker than that of a normal transistor. This thick-oxide transistor may increase the process cost but in many cases, there is a thick oxide transistor in a standard process for analog, memory and I/O blocks and they don't necessarily mean a cost increase. Moreover, since the thick-oxide power switch reduces not only the subthreshold leakage but also the tunneling leakage of a gate. Thus, it will be more meaningful when the development of high-k material is delayed.

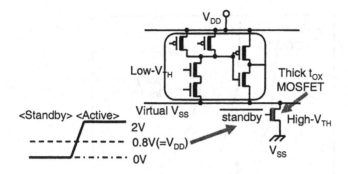

Figure 4-9. Boosted Gate MOS (BGMOS).

On the other hand, in Super Cut-off CMOS (SCCMOS) shown in Figure 4-10, the gate of a power switch is made negative in a standby mode. For example, if the gate voltage is set at -0.2V, the leakage current in a standby mode can be reduced about a factor of 1/100 due to the V_{GS} reverse bias. The power switch is made with a low-V_{TH} device so that the drivability is high even at the low-V_{DD} environments. Figure 4-11 shows a measured comparison among SCCMOS, MTCMOS and high-V_{TH} gate, all of which show the same leakage at a standby mode. SCCMOS clearly shows a superior speed at low V_{DD} and can be a power gating candidate for the future. Compared with BGMOS, SCCMOS can be made without thick oxide and thus has less process complication while handling negative voltage will give little more circuit complication. If readers are interested, the circuit schemes to handle the negative voltage are found in [13].

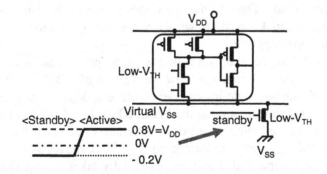

Figure 4-10. Super Cut-off CMOS (SCCMOS).

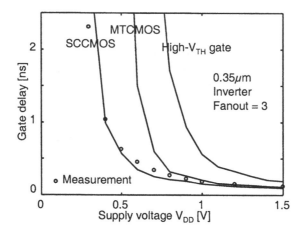

Figure 4-11. Measurement results for SCCMOS.

4.3.1.2 Power Gating with Quick Wake-up (Zigzag CMOS)

Currently one of the most effective ways to reduce dynamic power consumption is clock gating. In [12], it is reported that the clock gating alone reduced 70% of dynamic power in an MPEG4 chip. A bad news is that the clock gating loses its effectiveness in the leakage dominant era. Even though the clock is stopped, the subthreshold current continues to flow. It is easy to come up with the idea of using power gating instead of the clock gating. In the clock gating, however, the circuit should work from the next cycle when a wake-up signal comes. Thus, to use the power gating in the same context of clock gating, a quick wake-up within a fraction of a clock cycle is necessary for the power gating.

MTCMOS is taking several cycles in a wake-up process and can not be used instead of the clock gating. The slow recovery of MTCMOS is because the voltage of all internal logic nodes goes to close to V_{DD} when a standby mode lasts for a long time, assuming the MTCMOS uses a footer as a power switch. Then, before returning to a normal operation, almost a half of all internal nodes and V_{SSV} are discharged to V_{SS}. This means a full-swing discharge of a huge capacitance and this essentially takes a long time and even more time to make the peak current of the discharge within a practical limit.

In order to overcome the shortcoming of MTCMOS, the zigzag CMOS is introduced [13]. The circuit diagram and schematic waveforms for the zigzag CMOS are shown in Figure 4-12 and Figure 4-13. Before going into a standby, a input phase control (phase-forcing) is carried out so that '0' and '1' state of each internal node is always the same in a standby. For that purpose, before going into a standby, a phase-force signal is asserted and then, all internal node voltage are fixed irrespective of '0' and '1' of incoming signals to the logic block. A footer switch is inserted for the gate whose output is '1' after the phase forcing, and a header power switch is inserted for the gate whose output is '0' in a standby mode. Since this footer and header combination gives an impression of a zigzag, the scheme is called zigzag CMOS although it is not always exactly zigzag.

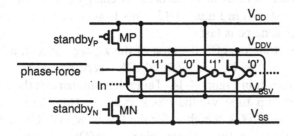

Figure 4-12. Zigzag SCCMOS.

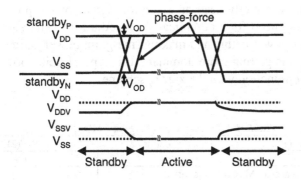

Figure 4-13. Schematic waveforms for zigzag SCCMOS.

The footer and header can be implemented as either MTCMOS, BGMOS or SCCMOS. The original proposal was using SCCMOS and it was called ZSCCMOS which corresponds to Figs. 4.12 and 4.13. In ZSCCMOS, after the phase-forcing, the gate of a footer is applied negative voltage and the gate of a header is applied the higher voltage than V_{DD}. Even if the gate of power switches is either V_{SS} or V_{DD} instead of voltage outside the power supply, one order of magnitude leakage reduction is possible due to the off-off MOSFET stacking effects where the DIBL (Drain Induced Barrier Lowering) effectively reduces the leakage. If V_{SS} and V_{DD} are used instead of negative voltage, the control circuit becomes very simple.

After the phase-forcing, '0' and '1' states do not change even though the standby mode lasts for a long time. This is because the gate whose output is '0' ('1') is directly connected to V_{SS} (V_{DD}) without any power switch in a standby mode. Consequently, in a wake-up process, there is no need for the discharging and charging of internal nodes and this enables a fast wake-up. Moreover, voltage of the virtual V_{DD} and V_{SS} lines change only about 0.1V ~ 0.2V in a standby mode and there is no need to charge and discharge the lines at wake-up as is shown in Figure 4-13. This is another reason why the wake-up of the zigzag scheme is fast.

In order to demonstrate the effectiveness of the zigzag concept, a Brent-Kung adder is built with the ZSCCMOS scheme [14], whose microphotograph is shown in Figure 4-14. Table 4.1 summarizes the results. The ratio of the wake-up time vs. the clock cycle is 16%. This number should be compared with 200% which is for the fastest MTCMOS wake-up time reported. The wake-up time of the zigzag CMOS is one order of magnitude shorter than that of MTCMOS and it is a fraction of a clock cycle. If the power gating is used as an alternate method of the clock gating, it is extremely important for the wake-up time to be much faster than a clock cycle time, because if this condition holds, the standby enable signal for the clock gating can be directly used as a control signal for the power gating. This eliminates the need for a complicated procedure for generating control timing signals for power gating and making fine-grain power gating practical. When the leakage is getting more dominant, this type of quick power gating will be used in conjunction with the zigzag scheme.

Table 4-1. Fast wake-up of zigzag CMOS.

Wake-up time(1)	Clock cycle time (2)	(1)/(2)
0.3ns	1.9ns	16%
Fastest wake-up MTCMOS reported (J. Tschanz et al. in the reference [15])		200%

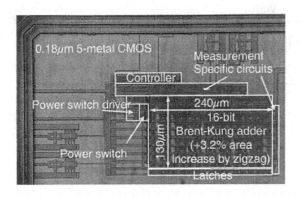

Figure 4-14. Test chip for zigzag CMOS.

4.3.2 Directions of commercial EDA tools for power gating

In order for the power gating technique to be widely accepted among designers, it is required to be supported by commercially available tools. However, EDA tools to solve issues specific to the power gating have not been supported so far. This situation is beginning to change. Sequence Design announced to support the MTCMOS power gating by offering the technology to analyze and optimize the design with the shared power switch [16]. In their approach, the power switch is encapsulated in and modeled as an independent cell, which is referred as a "switch cell". Low-Vth logic cells referred as "MTCMOS logic cells" share the virtual ground line and the switch cell, as depicted in Figure 4-15.

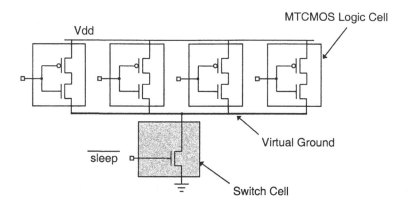

Figure 4-15. Cell-based architecture supported by a commercial EDA tool.

Design analysis and optimization for power gating are performed in two phases: the post-placement optimization and the post-route optimization. The design flow is shown in Figure 4-16. In the post-placement optimization for the power gating, the switch cells are automatically inserted to the optimal locations. Sizing the switch cells is performed as well taking into account the constraints. At the interface from the MTCMOS logic cell to a non-MTCMOS logic cell, a holder circuit is needed to prevent from floating. The holder circuit, referred as an "interface cell", is automatically inserted at the post-placement optimization stage as well.

In contrast, in the post-route optimization for the power gating, the voltage of the virtual ground is checked against the user-specified voltage limit. Resizing the switch cells is done when needed. These optimization capabilities for the power gating are incorporated into PhysicalStudio from Sequence Design.

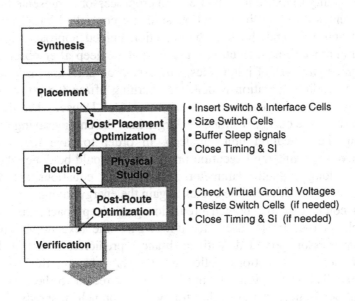

Figure 4-16. MTCMOS power-gating design flow.

4.3.3 Run-time power gating

Power gating techniques that have been developed so far mainly aim at reducing leakage current at the standby mode. However, the requirement for leakage reduction is extending to the active mode. This is because leakage power dissipation increases exponentially with device scaling and is projected to exceed dynamic power dissipation below 65nm feature size [17]. This means that leakage current becomes the major contribution to power dissipation in the active mode. As an attempt to save the leakage power in the active mode, run-time power gating is explored at various design levels. In this section, run-time power gating techniques at the architecture level and those at the gate level are presented.

4.3.3.1 Run-time power gating at architecture level

Run-time power gating is performed by detecting the idle periods of circuit components and dynamically turning off/on the power switches for the components. In [18], an architectural technique to apply the run-time power gating to execution units of a microprocessor is presented. Putting to sleep and waking up the execution units are controlled based on the time-based approach and the branch prediction guided approach. In the time-based approach, the execution units are put to sleep after observing a pre-determined number of idle cycles. The execution units are made to wake up once a pending operation is detected. Turning off/on the power switch for the entire execution unit incurs energy overhead. Hence, when the idle time is too short, the energy overhead is larger than the energy saving obtained by putting the execution unit to sleep. In order to save the total energy effectively, putting the execution unit to sleep should be done when the idle time is long enough. Parameterized analytical equations are derived to estimate the break-even idle cycles to gain the energy saving.

In contrast, in the branch prediction guided approach, the outcome of branch prediction is used to power-gate the execution units. In a microprocessor provided with a branch predictor, when a branch is mispredicted, instructions following the branch are flushed from the pipeline. The instruction fetching is then re-directed to the correct path of execution. Hence, in the cycles following a branch misprediction, a large part of the issue queue is flushed and the execution units are likely to be idle. In the branch prediction guided approach, when a branch misprediction is detected, the execution units are put to sleep immediately.

Analysis for the POWER4 microarchitecture indicates that using the time-based approach the floating-point units can be put to sleep for up to 28% of the execution cycles at a performance loss of 2% [18]. For the more difficult to power-gate fixed-point units, the branch prediction guided approach enables the fixed-point units to be put to sleep for up to 40% more of the execution cycles compared to the simpler time-based approach, with similar performance impact.

Architecture-level run-time power gating is also reported in [19]. Analytical energy models to determine the sleep-mode activation policies are studied for the integer functional units using dual-threshold domino logic circuits.

A compiler-assisted power-gating technique for functional units is discussed in [20]. The compiler identifies program regions in which functional units are expected to be idle and communicates this information to the hardware by issuing directives for turning units off at entry points of idle regions and directives for turning them back on at exits from such regions. Analysis shows that some of the functional units can be kept off for over 90% of the time at the performance degradation under 1%.

4.3.3.2 Run-time power gating at gate level

A run-time power gating approach to make use of existing gated-clock control signals is proposed in [21]. In this approach, the power switch for combinational logic gates is dynamically turned off to reduce leakage when their logic outputs are allowed to be "don't care" in the Finite-State-Machine (FSM) circuits.

Circuits implementing FSM are composed of state flip-flops (F/F's) to hold the state and combinational logic gates. In synchronous designs, data stored in the F/F's are updated in synchronous with the clock edge. Since dynamic power consumed in the clock network is quite large, a gated-clock technique [22] is commonly used to reduce dynamic power of clock portions. Figure 4-17 shows the circuit structure of gated-clock design. The clock signal "CLK" and the enable signal "EN" are ANDed to generate "Enabled_clock", being distributed to F/F's.

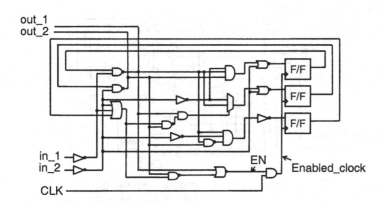

Figure 4-17. Circuit structure of gated-clock design.

When EN is 1, the clock signal CLK is transmitted to Enabled_clock and the output data of combinational logic network are loaded into the F/F's. When EN is 0, Enabled_clock stays 0, resulting in holding the data in the F/F's. Since Enabled_clock does not toggle at every cycle in contrast to CLK, dynamic power can be reduced. Thus, the gated-clock technique is used to reduce dynamic power at the clock network.

The enable signal in the gated clock, however, has the potential to reduce leakage if we look at the FSM circuit in the following way. In the gated-clock design, data stored in the F/F's are held when EN is 0. In other words, the data to be fed to the F/F's by the combinational logic network can be either 1 or 0 (i.e. don't care). In this condition, a portion of the combinational logic network is electrically disconnected from the ground to reduce leakage current. Figure 4-18 illustrates the "dynamic sleep control" scheme proposed in [21]. Combinational logic gates depicted as shaded in the figure are disconnected from the ground when EN is 0. Those combinational logic gates are referred as the "sleepable gates". A high-threshold power switch is provided in series with the sleepable gates and controlled by the signal EN. When EN is 1, the power switch turns on and the sleepable gates perform the normal logic operation. Results are stored into the F/F's. When EN is 0, since the power switch turns off the leakage current flowing through the sleepable gates is eliminated. The data in the F/F's is held irrespective of the output of the sleepable gates. Combinational logic gates to generate the enable signal are always made active because they are required to wake up the sleepable gates as well as to put them to sleep. Hence the combinational logic gates to generate the enable signal are referred as the "sleep control gates".

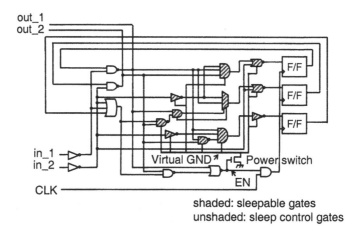

Figure 4-18. Dynamic sleep control scheme.

The algorithm to extract the sleepable gates from an FSM circuit is as follows. As a baseline, a circuit to which a gated-clock design is already applied is used. First, the enable signal EN of the gated clock is identified in the circuit. Subsequently, the logic network is traversed backwardly from EN toward inputs until reaching the primary inputs or the state F/F's. Every gate visited in the traversal is marked as a sleep control gate. The state F/F's and the AND gate for gated clock are also put into this set. Finally, the gates that do not belong to the set of the sleep control gates are identified as the sleepable gates.

Experiments using the MCNC benchmark circuits [23] demonstrate that 51-84% of logic gates are the sleepable gates in the FSM circuits. In addition, simulations for random input vectors indicate that those sleepable gates become idle in 53-90% of the time in the entire operation cycles except one example. Results are summarized in Table 4.2.

Table 4-2. Results from applying dynamic sleep control to MCNC benchmark.

Circuit name	% of sleepable gates	% of sleepable cycles
bbtas	54%	62%
bbara	68%	81%
bbsse	63%	53%
cse	63%	90%
dk14	67%	10%
keyb	84%	53%
lion9	51%	79%
sand	68%	56%
styr	72%	65%

Power analysis shows that leakage power in the active mode is reduced by 30-60%. Figure 4-19 shows the simulation results on the power dissipation. The entire power dissipation is saved by up to 20% at 125°C in 0.18μm technology.

Power saving at the high temperature alleviates problems in the burn-in testing such as throughput degradation and thermal runaway. In the burn-in, the chip is heated up typically to 100°C-140°C and tested in a burn-in oven. The total number of chips that can be simultaneously powered up for the burn-in testing will likely be limited by the maximum power dissipation capacity of the burn-in oven [24]. If all chips are active, then the total power dissipation can reach the several kilowatt range. The maximum number of chips that can simultaneously be powered up in burn-in is exponentially reduced as the technology scales [24]. The exponential increase in the leakage power is the main cause of this behavior. This leads to lowering the throughput of burn-in testing and increasing the test cost per chip. Thermal runaway is another problem that can be caused by increase of power dissipation at elevated temperature [25]. Increase of power dissipation leads to increasing heat dissipation. This results in raising the temperature and increasing the leakage current. Such positive feedback can eventually cause thermal runaway. Reducing leakage power in the active mode is essential for the burn-in testing at scaled devices.

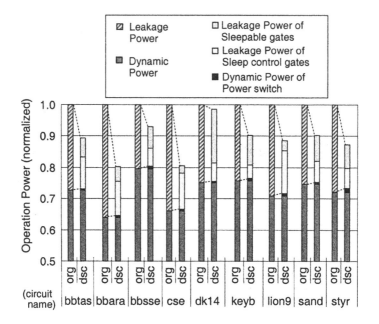

Figure 4-19. Comparisons on power dissipation between the original design (org) and the dynamic sleep control scheme (dsc) at 125°C, 1.8V, 50MHz in 0.18μm technology.

Timing issues in the dynamic sleep control scheme are discussed in [26]. It is reported that the critical path arises at the wake-up to raise the enable signal and put the sleepable gates back to operate. Analysis indicates that the critical path delay increases to 1.3X-1.8X compared to the non power-gated design. At the burn-in testing, however, a device is made to operate to activate internal transistors and hence is not necessarily made to run at the maximum frequency. In such a case this delay increase can be tolerated. By providing a scheme to switch between the BURN-IN-TEST mode and the NORMAL mode, the performance of normal operation is maintained. In the BURN-IN-TEST mode the power switch is dynamically turned off/on, while in the NORMAL mode the power switch is kept on. In order to apply the dynamic sleep control scheme to the NORMAL mode as well, the methodology to speed up the critical path for the wake-up needs to be developed.

4.4 SUMMARY

Power gating is a very effective technique to reduce the leakage power in nanoscale CMOS circuits. However, there are many issues to be solved when applying to real designs. In the scheme to share the power switch, verification and validation of the critical-path delay is difficult because the delay is sensitive to input vectors. Sizing the power switch plays an important role. By securing the sufficient power-switch size, the delay becomes insensitive to input vectors. Based on this fact, the required minimum size of the power switch is determined from the average current of the entire circuit. Unsharing the power switch gives another solution to avoid the complex timing problem. By providing a power switch for individual cell and selectively using such cells only in the critical paths, the performance is improved at the minimum area penalty. In order for the power gating technique to be widely accepted among designers, the support by commercially available tools is required. In addition, in order to be employed in future devices, circuit techniques for power gating at the scaled supply voltage becomes necessary. Methodologies for run-time power gating to reduce leakage power in the active mode will be essential as well.

REFERENCES

[1] S. Mutoh, T. Douseki, Y. Matsuya, T. Aoki, J. Yamada, "1V high-speed digital circuit technology with 0.5µm multi-threshold CMOS", *Proc. Sixth Annual IEEE International ASIC Conference and Exhibit*, pp. 186 - 189, September 1993.

[2] J. Kao, A. Chandrakasan, D. Antoniadis, "Transistor sizing issues and tool for multi-threshold CMOS technology," *Proc. 34th Design Automation Conf.*, pp. 409-414, June 1997.

[3] S. Mutoh, S. Shigematsu, Y. Gotoh, S. Konaka, "Design method of MTCMOS power switch for low-voltage high-speed LSIs," *Proc. IEEE Asia and South Pacific Design Automation Conf. (ASP-DAC'99)*, pp. 113-116, Jan. 1999.

[4] H. Won, K. Kim, K. Jeong, K. Park, K. Choi, J. Kong, "An MTCMOS design methodology and its application to mobile computing," *Proc. 2003 Int. Symp. on Low Power Electronics and Design*, pp. , Aug. 2003.

[5] K. Usami, N. Kawabe, M. Koizumi, K. Seta, T. Furusawa, "Automated selective multi-threshold design for ultra-low standby applications," *Proc. 2002 Int. Symp. on Low Power Electronics and Design*, pp. 202-206, Aug. 2002.

[6] S. Shigematsu, S. Mutoh, Y. Matsuya, Y. Tanabe, J. Yamada, "A 1-V high-speed MTCMOS circuit scheme for power-down application circuits," *IEEE J. of Solid-State Circuits*, vol. 32, no. 6, pp. 861-869, June 1997.

[7] K. Usami, N. Kawabe, M. Koizumi, K. Seta, T. Furusawa, "Selective multi-threshold technique for high-performance and low-standby applications," *IEICE Trans. on Fundamentals of Electronics, Communications and Computer Sciences*, vol. E85-A, no.12, pp. 2667-2773, Dec. 2002.

[8] K. Usami, M. Igarashi, F. Minami, T. Ishikawa, M. Kanazawa, M. Ichida, and K. Nogami, "Automated low-power technique exploiting multiple supply voltages applied to a media processor", *IEEE J. Solid-State Circuits*, vol.33, no.3, pp. 463-472, March 1998.

[9] G. Uvieghara, et al, "A highly-integrated 3G CDMA2000 1X cellular baseband chip with GSM/AMPS/GPS/Bluetooth/multimedia capabilities and ZIF RIF support", *2004 ISSCC, Digest of Tech. Papers*, vol.1, pp. 422-536, Feb. 2004.

[10] T. Inukai, M. Takamiya, K. Nose, H. Kawaguchi, T. Hiramoto, T. Sakurai, "Boosted gate MOS (BGMOS): device/circuit cooperation scheme to achieve leakage-free giga-scale integration," *CICC'00*, p.409, May 2000.

[11] H. Kawaguchi, K. Nose, T. Sakurai, "A CMOS scheme for 0.5V supply voltage with pico-ampere standby current," *1998 ISSCC, Digest of Tech. Papers*, pp.192-193, Feb. 1998.

[12] M. Ohashi, et al., "A 27MHz 11.1mW MPEG- 4 video decoder LSI for mobile application," *ISSCC*, pp.366-367, Feb.2002.

[13] K. S. Min, T. Sakurai, "Zigzag super cut-off CMOS (ZSCCMOS) block activation with self-adaptive voltage level controller: An alternative to clock-gating scheme in leakage dominant era," *ISSCC*, pp.400-401, Feb. 2003.

[14] T. Miyazaki, T. Q. Canh, H. Kawaguchi, T. Sakurai, "Observation of one-fifth-of-a-clock wake-up time of power-gated circuit," *CICC'04*, paper#6-1, Oct. 2004.

[15] J.Tschanz, S.Narendra, Y.Ye, B.Bloechel, S.Borkar, "Dynamic sleep transistor and body bias for active leakage power control of microprocessors," *IEEE J. Solid-State Circuits*, vol.38, pp.1838-1843, Nov. 2003.

[16] Sequence Design, Inc., "Leakage power solutions," *NanoCool Low-Power Design Seminar*, Tokyo, Nov. 10, 2004.

[17] N. Kim, T. Austin, D. Blaauw, T. Mudge, K. Flautner, J. Hu, M. J. Irwin, M. Kandemir and V. Narayanan, "Leakage Current: Moore's Law Meets Static Power," *IEEE Computer*, vol.36, no.12, pp.68-75, December 2003.

[18] Z. Hu, A. Buyuktosunoglu, V. Srinivasan, V. Zyuban, H. Jacobson, P. Bose, "Microarchitectural techniques for power gating of execution units," *Proc. 2004 Int. Symp. on Low Power Electronics and Design*, pp. 32-37, Aug. 2004.

[19] S. Dropsho, V. Kursun, D. Albonesi, S. Dwarkadas, E. Friedman, "Managing static leakage energy in microprocessor functional units," *Proc. 35th Annual IEEE/ACM Int. Symp. on Microarchitecture (MICRO-35)*, pp. 321- 332, Nov. 2002.

[20] S. Rele, S. Pande, S. Onder, R. Gupta, "Optimizing static power dissipation by functional units in superscalar processor," *Proc. Int. Conf. on Compiler Construction*, pp. 261-275, Apr. 2002.

[21] K. Usami and H. Yoshioka, "A scheme to reduce active leakage power by detecting state transitions," *Proc. 47th IEEE Int. Midwest Symp. on Circuit and Systems (MWSCAS'04)*, I493-I496, July 2004.

[22] L. Benini and G. De Micheli, *Dynamic Power Management: Design Techniques and CAD Tools*, Kluwer Academic Publishers, 1998.

[23] [Online] http://www.cbl.ncsu.edu/

[24] O. Semenov, A.Vassighi, M. Sachdev, A. Keshavarzi and C. Hawkins, "Effect of CMOS technology scaling on thermal management during burn-in," *IEEE Trans. Semiconductor Manufacturing*, vol.16, no.4, Nov. 2003.

[25] A.Vassighi, O. Semenov, M. Sachdev, A. Keshavarzi, "Effect of static power dissipation in burn-in environment on yield of VLSI," *Proc. IEEE Int. Symposium on Defect and Fault Tolerance in VLSI Systems (DFT'02)*, 2002.

[26] K. Usami and H. Yoshioka, "Dynamic sleep control for finite-state-machines to reduce active leakage power," *IEICE Trans. on Fundamentals of Electronics, Communications and Computer Sciences*, vol. E87-A, no.12, pp. 3116-3123, Dec. 2004.

Chapter 5

BODY BIASING

Tadahiro Kuroda[§] and Takayasu Sakurai[¶]
§*Keio University, Japan and* ¶*University of Tokyo, Japan*

5.1 INTRODUCTION

Reverse body biasing has been widely used in commercial memory chips since the mid-1970s, in order to lower the risk of latchup and memory data destruction, due to lack of substrate contacts for high density cell layout. In logic chips, on the other hand, the substrate and wells are typically biased stably to the ground and power potential with sufficient substrate contacts to ensure that no devices become forward biased, raising the risk of latchup due to unexpected operation of random logic circuits. Since the mid-1990s, however, reverse body biasing has been applied in logic chips for a different reason: power reduction.

CMOS power dissipation is increasing rapidly by device scaling [1]. Lowering power supply voltage, V_{DD}, is effective in reducing the power dissipation, but at a cost of increase in propagation delay time. In order to recover the circuit speed, transistor threshold voltage, V_{TH}, should be lowered [2, 3, 4]. This approach, however, raises two problems.

The first problem is rapid increase in sub-threshold leakage in low V_{TH} devices. For every 0.1-volt reduction of V_{TH}, sub-threshold leakage current increases by about one decade. Battery life in portable equipment shortens unless this leakage current is reduced in a standby mode. In standby leakage current (IDDQ) testing, it is difficult to sort out defective chips by monitoring the quiescent power supply current, because leakage current caused by a defect cannot be detected under cover of the increased sub-threshold leakage current. Since some kinds of defects are difficult to detect by means other than the IDDQ testing [5], defect mixed rate may be increased. Without the IDDQ testing it is more difficult to develop test vectors for high test coverage as integration level improves. It is also expensive to provide large enough current to chips in a burn-in testing, since V_{TH} is lowered at high temperature to further increase sub-threshold leakage.

The second problem is degradation of worst-case speed due to V_{TH} variation in low voltage operation [6, 7]. Delay variation due to V_{TH} variation, ΔV_{TH}, is substantially increased at low V_{DD}'s. The increased variation of the operating speed degrades the chip performance, or increases the chip size and the power dissipation to compensate for the speed degradation. In order to keep the delay variation percentage constant in low V_{DD}'s, ΔV_{TH} should be reduced approximately by [8]

$$\frac{\Delta V_{TH}'}{\Delta V_{TH}} = \left(\frac{T_{pd}}{T_{pd}'} \cdot \frac{V_{DD}'}{V_{DD}} \right)^{\frac{1}{1.4}}$$

(5.1)

where T_{pd} is propagation delay time. For example, when V_{DD} is lowered from 1.5V to 1.0V and V_{TH} is lowered to maintain circuit speed (i.e. $T_{pd}=T_{pd}'$) , ΔV_{TH} should be reduced by 18%. It is very difficult to lower ΔV_{TH} by this much by means of process and device refinement.

Power gating either on a board or in a chip can solve the battery life problem, as discussed in Chapter 3 and 4, but it cannot solve the other problems in association with the IDDQ testing, the burn-in testing, and the speed degradation due to V_{TH} variation. Substrate biasing can solve all these problems, since designer can control V_{TH} by utilizing the body effect. In the active mode V_{TH} is compensated for process variation and accurately set to a low voltage for low-power, high-speed circuit operation at low supply voltages. In the standby mode, the IDDQ testing, and the burn-in testing, V_{TH} is raised so that the sub-threshold leakage can be reduced substantially.

The body effect is given by the following equation.

$$V_{TH} = V_{TH0} + \gamma \left[\sqrt{(2\phi_b - V_{BS})} - \sqrt{2\phi_b} \right],$$

$$\gamma = \frac{t_{ox}}{\varepsilon_{ox}} \sqrt{2\varepsilon_{si}qN_A} , \quad \phi_b = \frac{kT}{q} \ln\left(\frac{N_A}{N_i} \right) ,$$

(5.2)

in which V_{BS} is the substrate potential, V_{TH0} is V_{TH} for V_{BS} =0V, t_{ox} is the gate-oxide thickness, ε_{ox} is the dielectric constant of the silicon dioxide, ε_{si} is the permittivity of silicon, N_A is the doping concentration density of the substrate, N_i is the carrier concentration in intrinsic silicon, k is Boltzmann's constant, q is the electronic charge, and T is the absolute temperature. The constant, γ, characterizes the body effect, and is called the body effect coefficient.

In recent years, the purpose of substrate biasing has been extending. Originally, it was utilized to reduce sub-threshold leakage in a standby mode for portable applications and in testing. More recently, it has been employed to reduce the maximum power dissipation by lowering V_{TH} in active mode, and by compensating variations in V_{TH} in a high-end microprocessor. In the future, it may be used for extending battery life in an active mode by adjusting V_{TH} in accordance with workload.

Accordingly, the bias range has extended from reverse body bias (RBB) to forward body bias (FBB) in order to maintain the coverage of V_{TH} change in scaled CMOS devices. Adaptive control techniques are also changing in terms of a control scheme (closed loop control or open loop control), control objectives (V_{TH} alone, or together with V_{DD} and frequency), a control method (analog control, digital control, software control), and accuracy (space granularity and time granularity).

In this chapter, circuit techniques as well as control schemes for substrate biasing will be presented and future directions will be discussed. Please note that the terms, **body bias**, **substrate bias**, **well bias**, and **backgate bias** refer to the same concept of the threshold voltage modulation using the fourth terminal of the MOSFET.

5.2 REVERSE BODY BIAS

5.2.1 Variable Threshold-voltage CMOS (VTCMOS) Technology

A VTCMOS technology [7, 8, 9, 10, 11, 12] controls V_{TH} by means of substrate bias control. In this technique, devices are fabricated for lower V_{TH} than a design target, and V_{TH} is set to the target by adjusting the substrate bias, V_{BB}. Since sub-threshold leakage current depends very strongly on V_{TH}, V_{TH} can be compensated for variations by feedback control of V_{BB} such that monitored leakage current is set to a target value. In the standby mode, the IDDQ testing, and the burn-in testing, the bias is adjusted to set the target V_{TH} value high enough to reduce the leakage current by several decades as compared to active mode. Three schemes are developed for different purposes.

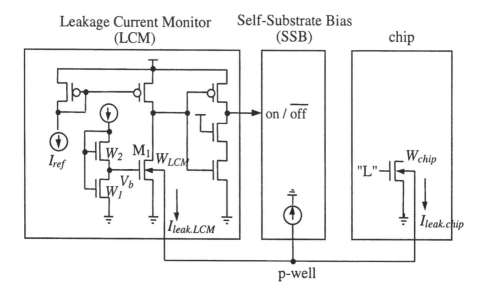

Figure 5-1. Self-Adjusting Threshold-voltage (SAT) scheme.

5.2.1.1 Self-Adjusting Threshold-voltage (SAT) Scheme

An SAT scheme, depicted in Figure 5-1, compensates for the V_{TH} variation [8, 9]. The sub-threshold leakage current is monitored by a Leakage Current Monitor (LCM). The substrate bias is generated by a Self

Substrate Bias circuit (SSB). LCM activates SSB when a monitored leakage current in LCM, $I_{leak.LCM}$, is larger than a target preset value, I_{ref}. SSB lowers V_{BB} by pumping out current from the substrate [13]. Accordingly, V_{TH} is raised and $I_{leak.LCM}$ is reduced. When $I_{leak.LCM}$ becomes smaller than I_{ref}, LCM stops SSB. However, the substrate current due to the impact ionization and the junction leakage raises V_{BB} gradually again. Accordingly, V_{TH} is lowered gradually and $I_{leak.LCM}$ increases. When $I_{leak.LCM}$ becomes larger than I_{ref}, LCM activates SSB again. By activating SSB intermittently in this way, V_{TH} can be set to the target value, and consequently, its process induced variation can be compensated to be smaller. Since the SAT scheme can compensate for the V_{TH} variation but cannot reduce standby power dissipation, it is suitable for active power reduction and circuit speed improvement of LSI's for desktop use.

5.2.1.2 Standby Power Reduction (SPR) Scheme

An SPR scheme is for reducing the sub-threshold leakage current in the standby mode as well as in the IDDQ testing and the burn-in testing [10, 11].

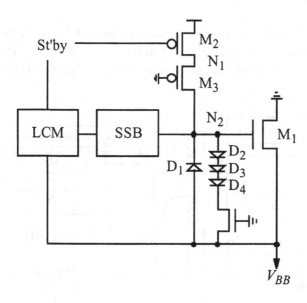

Figure 5-2. Standby Power Reduction with SSB (SPR w/SSB) scheme.

In the active mode, the p-well voltage is set to V_{SS} and the n-well to V_{DD} (zero body bias; ZBB). In the standby mode and the IDDQ testing, the reverse body bias (RBB) is applied by using external power supplies or SSB, in order to raise V_{TH}.

There are three implementations. The first one is to connect the substrate contacts to pads so that RBB is applied in the testing by external power supplies for the leakage current reduction. In shipping the chip as a product, V_{SS} and V_{DD} are provided to the Pads for ZBB.

The second implementation is to switch the substrate bias between RBB and ZBB by using a switch circuit, namely an SPR with Switch scheme [10]. The advantages of this approach are that the circuit is very simple and small and that switching can be made faster than 0.1μs. The drawback, on the other hand, is that two additional external power supplies are required.

The third implementation, depicted in Figure 5-2, is to generate the substrate bias in the standby mode by SSB, namely an SPR w/ SSB scheme [8]. The advantage of this approach is that the standby power reduction can be carried out without adding any external power supply. The drawback is that it takes several hundred microseconds to reduce the leakage current, even though it takes only 0.1μs to return to the active mode.

Since the SPR scheme reduces sub-threshold leakage in the testing and the standby mode but cannot compensate for the V_{TH} variation, it is suitable for maintaining test quality and power reduction of LSI's for portable equipment when V_{TH} is lowered for high speed operation at low voltages.

5.2.1.3 SAT+SPR Scheme

The third scheme is a combination of the SAT and SPR schemes, depicted in Figure 5-3 [8]. The substrate is biased deeper by SSB and shallower by a Substrate Charge Injector (SCI). Figure 5-4 illustrates the

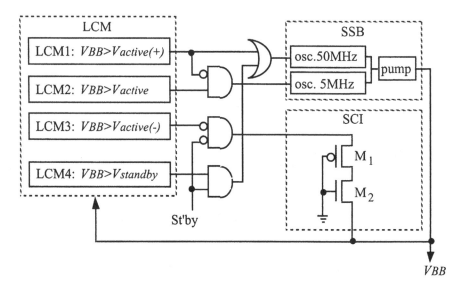

Figure 5-3. SAT+SPR scheme.

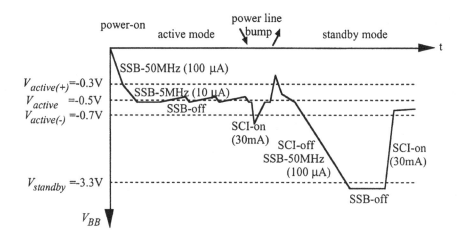

Figure 5-4. Substrate bias control in SAT+SPR scheme.

substrate bias control. By setting two additional substrate potentials as references, $V_{active(+)}$ and $V_{active(-)}$, around a target reference of V_{active}, recovery time is shortened when V_{BB} moves away from V_{active}, and also controllability around V_{active} is improved. After a power-on, V_{BB} is higher than $V_{active(+)}$, and SSB begins to draw 100μA from the substrate to lower V_{BB} using a 50MHz ring oscillator. This current is large enough for V_{BB} to settle down within 10μs after a power-on. When V_{BB} goes lower than $V_{active(+)}$, the pump driving frequency drops to 5MHz and SSB draws 10μA to control V_{BB} more precisely. SSB stops when V_{BB} drops below V_{active}. V_{BB}, however, rises gradually due to device leakage current through MOS transistors and junctions, and reaches V_{active} to activate SSB again. In this way, V_{BB} is controlled at V_{active} by the on-off control of the SSB. When V_{BB} goes deeper than $V_{active(-)}$, SCI turns on to inject 30mA into the substrate. Therefore, even if V_{BB} jumps beyond $V_{active(+)}$ or $V_{active(-)}$ due to power line bump for example, V_{BB} is quickly recovered to V_{active} by SSB and SCI. When the St'by signal is asserted to go to the standby mode, SCI is disabled and SSB is activated again and 100μA current is drawn from the substrate until V_{BB} reaches $V_{standby}$. V_{BB} is set at $V_{standby}$ in the same way by the on-off control of the SSB. When the St'by signal becomes "0" to go back to the active mode, SSB is disabled and SCI is activated. SCI injects 30mA current into the substrate until V_{BB} reaches $V_{active(-)}$. V_{BB} is finally set at V_{active}. In this way, SSB is mainly used for transition from the active to the standby mode, whereas SCI is used for transition from the standby to the active mode. The active-to-standby mode transition takes about 100μs, while the standby-to-active mode transition is completed in 0.1μs.

This scheme is suitable for power reduction, circuit speed improvement, and test quality maintenance of LSI's for portable use.

5.2.2 VTCMOS Circuit Techniques

Three circuit techniques are essential: (1) the leakage current monitor, LCM, (2) the self-substrate bias generator, SSB, and (3) a combination of SSB and a charge injector.

5.2.2.1 Leakage Current Monitor

In Figure 5-1, the ratio of $I_{leak.LCM}$ to the total leakage current in a chip, $I_{leak.chip}$, or the leakage current detection ratio, X_{LCM}, is given by

$$X_{LCM} \equiv \frac{I_{leak.LCM}}{I_{leak.chip}} = \frac{W_{LCM} \, 10^{(V_b - V_{TH})/S}}{W_{chip} \, 10^{-V_{TH}/S}} = \frac{W_{LCM}}{W_{chip}} \cdot 10^{\frac{V_b}{S}} , \qquad (5.3)$$

where W_{chip} is effective total channel width corresponding to the total leakage current in the chip, W_{LCM} is channel width of a monitor transistor in LCM, and V_b is its gate potential. Since $I_{leak.LCM}$ leads to a power penalty of LCM it should be as small as possible. Too small $I_{leak.LCM}$, however, slows LCM response speed, which enlarges fluctuation of V_{BB} caused by the on-off control of SSB, resulting in larger dynamic error of V_{TH}. When $I_{leak.LCM}$ is 1μA for the chip leakage current of 1mA, X_{LCM} is 0.1%. Given $V_b = 2S$, which is approximately 0.2 volts, the size of the monitor transistor can be designed as small as approximately 0.001% of the effective total transistors in the chip.

A bias circuit for V_b that is depicted in Figure 5-1 is developed. A current source is designed such that the two transistors are operated in the sub-threshold region. As the drain currents of the two transistors are equal,

$$W_2 \cdot 10^{(V_1 - V_b - V_{TH})/S} = W_1 \cdot 10^{(V_1 - V_{TH})/S} ,$$

$$\therefore \ V_b = s \cdot \log \frac{W_2}{W_1} . \qquad (5.4)$$

Substituting Eq.(5.4) into Eq.(5.3),

$$X_{LCM} = \frac{W_{LCM}}{W_{chip}} \cdot \frac{W_2}{W_1} . \qquad (5.5)$$

X_{LCM} can be designed only by transistor size ratio and independent of the power supply voltage, temperature, and process variation. If V_b is generated by dividing voltages between V_{DD} and V_{SS} by resistors, $V_b = \lambda \cdot V_{DD}$, and consequently, X_{LCM} is a function of V_{DD} and S. Since S is a function of temperature, X_{LCM} depends on V_{DD} and temperature, which is not desirable. Variation in X_{LCM}, analyzed by SPICE simulation, is within 15%, which results in less than 1% error in V_{TH} controllability.

5.2.2.2 Self-Substrate Bias

A schematic diagram of a pump circuit in SSB is depicted in Figure 5-5. P-channel transistors of the diode configuration are connected in series and their intermediate nodes are driven by two signals, Φ1 and Φ2, in 180° phase

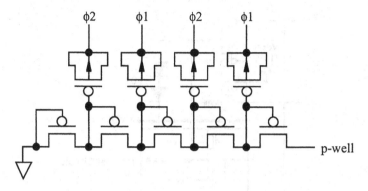

Figure 5-5. Pump circuit.

shift. Every other transistor therefore sends current alternately from p-well to V_{SS}, resulting in lower p-well bias than V_{SS}. The SSB circuits are widely used in DRAM's and E^{2}PROM's. The driving current of SSB is around 100μA. Circuit size can be smaller at low V_{DD} since substrate current generation due to the impact ionization is reduced significantly at low supply voltages.

5.2.2.3 Combination of SSB and Charge Injector

When using both SSB and a charge injector such as M_1 in Figure 5-2 and SCI in Figure 5-3, special care should be paid to keep any transistor from receiving high voltages and to prevent leakage current.

In the active mode in Figure 5-2, SSB stops, and transistor M_1 is turned on, and the substrate is connected to V_{SS}. In the standby mode, M_1 is turned off, and the substrate bias is controlled by LCM and SSB in the same way as in the SAT scheme. In order to turn off M_1 completely, the gate potential should be lowered with respect to the substrate potential, as the source is connected to the substrate. For this purpose, a diode D_1 is inserted between the gate and the source of M_1, and SSB is connected to the gate. When SSB pumps current out and the gate potential of M_1 reaches -0.7 volts, the diode D_1 turns on. Thereafter SSB keeps the gate-source voltage of -0.7 volts to turn M_1 off completely, and pumps the substrate current out to lower the substrate potential. At this point, in order to keep a transistor M_2 from receiving over-voltage, a transistor M_3 whose gate is connected to V_{SS} is inserted between the drain of M_2 and the gate of M_1 to clamp the drain potential of M_2 higher than $V_{SS}+V_{TH}$. In transiting to the active mode, both LCM and SSB are disabled, M_2 is turned on and so is M_1. At this point, the gate voltage of M_1 should be raised gradually in accordance with the rise in

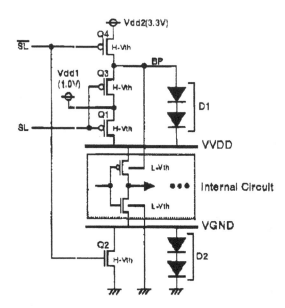

Figure 5-6. Auto-Backgate-Controlled MT-CMOS circuit.

the substrate potential for the reliability. By inserting three diodes, D_2, D_3, and D_4, between the gate and the source, the gate-source voltage of M_1 is clamped to 1.8 volts for protection. When the substrate potential rises to V_{SS}-V_{TH}, this clamp circuit is released, and V_{DD} is provided to the gate to turn on M_1 strongly.

In SCI in Figure 5-3, a p-channel transistor M_1 whose gate is connected to V_{SS} is inserted to divide the voltage of $V_{DD}+|V_{BB}|$ and keep any transistor from receiving high voltages. Furthermore, to keep the substrate from being biased forwardly, an n-channel transistor M_2 whose gate is connected to V_{SS} is also inserted. $|V_{GS}|$ and $|V_{GD}|$ of M_1 and M_2 never exceeds the larger of V_{DD} and $|V_{standby}|$.

5.2.2.4 Combination of RBB and Power Gating

RBB can be applied by raising source potential, instead of applying negative potential to body. An example circuit is depicted in Figure 5-6 [14]. In the standby mode when power gating transistors (Q1 and Q2) are turned off, the potential voltage of virtual power line VGND is raised and VDD is lowered by diodes (D1 and D2) to the point where chip leakage current and diode forward current balance. Body potential is determined by the number of the diode connections in series and the size of them. It is an advantage that no pump circuit is required. It is a drawback, though, that fine control of V_{TH} is very difficult.

A 32-bit RISC microprocessor is implemented with the body bias control circuit in Figure 5-6 [14]. Measured threshold voltages of NMOS transistors are 0.12V and 0.40V, and those of PMOS devices are -0.20V and -0.48V, in the active mode and the standby mode, respectively.

5.2.3 VTCMOS Performance and Penalty

5.2.3.1 V_{TH} Controllability

An MPEG-4 video codec chip [15] is fabricated in two runs. The target of V_{TH} in one run is 0.05V and that for the other is 0.15V by changing conditions of ion-implantation. About 40 chips are measured for each V_{TH} condition in the following three ways: (1) V_{TH} as processed by ZBB; (2) V_{TH} controlled by VTCMOS in the active mode, and (3) V_{TH} controlled by VTCMOS in the standby mode. In (2), the MPEG-4 chip is operated with test vector inputs so that the measurements include dynamic errors, such as those due to substrate noise influence. The measured results at 27°C and 70°C are plotted in Figs. 5.7 (a)-(d). The VTCMOS technology reduces V_{TH} variation from ±0.1V to ±0.05V in both the active and the standby modes, and raises V_{TH} by 0.25V in the standby mode.

Measured temperature dependence of V_{TH} is 0.7mV/°C for an NMOS and -0.7mV/°C for a PMOS under the VTCMOS control, whereas the values are -1.3mV/°C and 2.0mV/°C, respectively, in the conventional CMOS device. When V_{DD} is around 0.5V, the drain current shows positive temperature dependence, since the increase in the drain current by V_{TH} decrease surmounts the mobility degradation [16]. This may cause thermal runaway if the sub-threshold leakage becomes the dominant component in power dissipation at low V_{TH}. Thus, in a scaled device with low V_{DD} and low V_{TH}, temperature dependence control becomes indispensable. The temperature dependence of V_{TH} in VTCMOS can be controlled by controlling the temperature dependence of the target leakage current source (I_{ref}) in LCM.

Chip leakage current is measured at 27°C and 70°C, and the results are plotted in Figure 5-8. The horizontal axes is the average of $|V_{TH.p}|+V_{TH.n}$. The VTCMOS technology sets the leakage current below 10mA in the active mode and below 10μA in the standby mode, independently from processed V_{TH} and temperature.

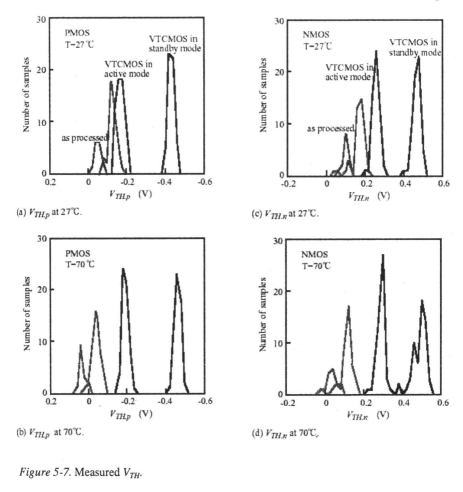

Figure 5-7. Measured V_{TH}.

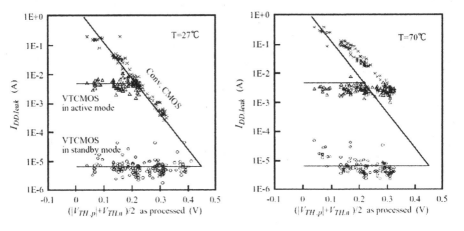

Figure 5-8. Measured chip leakage current.

5.2.3.2 Switching Response

In the SPR w/ SSB scheme and the SAT+SPR scheme it takes 100µs for SSB to pump current out to transit to the standby mode, whereas it takes 0.1µs for a transistor to inject current to go to the active mode. This "slow falling asleep but quick awakening" feature is in most cases acceptable.

5.2.3.3 Power Penalty

Power penalty for SSB and LCM is very small. Substrate current due to impact ionization is approximately only a few tenths of a percent of the power current. The substrate current is almost independent of the operating frequency. From this fact, it is inferred that current due to the impact ionization at non-switching transistors dominates the substrate current at low V_{TH}. SSB in its equilibrium control pumps out the same amount of the substrate current as is generated, for which SSB consumes several times more current from the power supply. Consequently, SSB power dissipation is less than 1% of the chip power current, when the substrate potential is under equilibrium control. LCM, on the other hand, consumes power all the time to monitor the leakage current. The smaller the power for LCM, the slower the response time, and the larger the control error of V_{BB}. Typical value of the LCM current is 3µA. In the standby mode, this static power is larger than the dynamic power of SSB. In most systems, however, smaller than 10µA standby current is considered negligible.

Switching between the active mode and the standby mode, by charging and discharging substrate capacitance, also consumes power. Suppose, for instance, that the substrate capacitance for 10mm^2 is 5nF, and that the substrate potential is changed by 3 volts. The energy required for charging and discharging the substrate each time is 5nF x (3V)^2, that is 0.05 µjoule, and is negligibly small.

5.2.3.4 Area Penalty

Compared with the power gating where the power supply is controlled, the substrate bias can be controlled by a much smaller circuit, since much smaller current flows in the substrate than in the power supply lines. Area penalty of the VTCMOS circuit itself is less than 1%, but area penalty for separation of the substrate contacts may be 6% at most. In order to bias the substrate all the substrate contacts should be routed together to the substrate bias control circuits. A practical layout method is illustrated in Figure 5-9. Substrate contacts in a leaf cell are removed automatically by a mask data processing program so that a standard cell library does not have to be re-designed. Substrate contacts are automatically placed with regular vertical lines for distributing substrate biasing. The area penalty can be reduced if the number of the substrate distributing lines is reduced, which raises concerns about the latch-up immunity and the substrate noise influence.

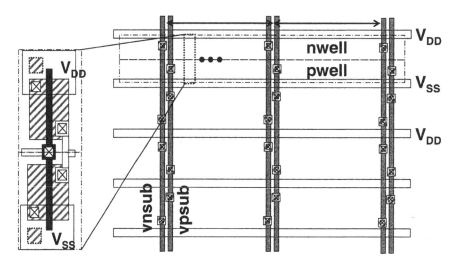

Figure 5-9. Substrate contacts layout in VTCMOS.

5.2.3.5 Latch-Up Immunity

Generally the following three triggers are considered to initiate latchup: (1) forward bias of the substrate with respect to the source diffusions during power-up due to capacitance coupling between them, (2) substrate current due to impact ionization in the active mode and the burn-in test, and (3) current injection from the I/O circuits due to overshoot and undershoot of input and output signals.

While the SSB is not sufficiently operated during power-up, the substrate may be biased forwardly due to capacitance coupling between the source and the substrate, resulting in rush power current or latchup problems [17]. A parasitic lateral npn-bipolar transistor, Q_L, is first turned on. The collector current of Q_L in turn lowers the potential of the n-well against the p-channel transistor source, and turns on a parasitic vertical pnp-bipolar transistor, Q_V, to initiate latchup. The same thing may happen in DRAM. However, in DRAM the n-well is tied to V_{DD} and only the p-substrate is biased by SSB. In this case, Q_L is more likely to turn on than in VTCMOS, but Q_V is more unlikely to turn on because the n-well is tied to V_{DD}. Since VTCMOS biases both the p-substrate and the n-well, latch-up immunity during power-up may be degraded in VTCMOS. However, latch-up prevention techniques developed for DRAM can be applied to VTCMOS. For example, during power-up a power-on-shunt circuit works to shunt the p-substrate to V_{SS} and the n-well to V_{DD} [17]. The power-on-shunt circuit is depicted in Figure 5-

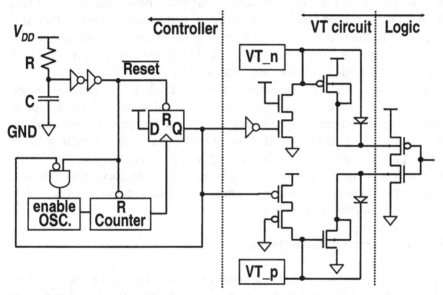

Figure 5-10. Power-on-shunt circuit to prevent latch-up during power-up.

10. It works in the following way: a connection node of a resistor (R) and a capacitor (C) in series between V_{DD} and V_{SS} first resets a D-type flip-flop and initializes a counter, and then it activates the counter after the time constant of RC. When the counter counts up to a certain number the flip-flop is set and the counter stops. The Q output of the flip-flop is used for the power-on-shunt signal which shunts the substrate to V_{SS} for that time after power-up.

Internal latchup in the active mode can be prevented by designing high current pumping capability for the SSB. In the burn-in test at over-voltages the substrate current due to impact ionization may exceed current pumping capability of the substrate bias control circuit, and finally the substrate bias rises abruptly, thereby causing latch-up or a latchup-like breakdown phenomenon [17]. Therefore the SSB must be designed to handle the substrate current not only during normal operation, but also during the burn-in test.

I/O latch-up is reduced, since the substrate is not tied to the power lines. For example, when undershoot signal of $-V_{in}$ ($<-0.7V$) is received at an input, current flows out from the p-substrate through an input ESD protection device. In this case, the p-substrate is biased to $-V_{in}+0.7V$ ($<0V$), due to a pn-junction between p-substrate and n-diffusion in the ESD protection device, and therefore Q_L is unlikely to turn on. When overshoot signal is received, Q_V is unlikely to turn on in the same way. However, when another n-well exists close by, Q_V may eventually turn on there. One way to prevent it is to apply high bias to the n-well for the input protection device to keep Q_V from being turned on by the overshoot signal. Another sure method is to employ a triple-well structure to bias only the internal p-well and n-well, which are completely separated from the I/O portion by a deep n-well. In this approach, high V_{TH} is also necessary for the I/O circuits.

In any case, when V_{DD} becomes lower than the holding voltage (typically around 1.5V), latch-up never occurs. The substrate resistance can be reduced by employing an epitaxial layer or a retrograded well, and the current gain of Q_L can be lowered by introducing a shallow trench isolation. These process options raise the holding voltage. As long as V_{DD} for the burn-in test is higher than the holding voltage, latch-up prevention should be considered in circuit and layout designs. Eventually, however, low-voltage CMOS will be free from latch-up.

5.2.3.6 Substrate Noise Influence

Tracking jitter for DLL and a shmoo plot for SRAM, both of which are very sensitive to substrate noise, are measured under noise generated on the same chip[12]. The DLL and the SRAM share the same p- and n- wells with the noise generator. Two variants are designed in terms of the number of the substrate contacts per the well strip: (1) only one substrate contact at the bottom, and (2) 400 substrate contacts in equal density. In (1), the noise generator is operated at 50MHz, whereas, in (2), the noise generator is disabled. Measured DLL tracking jitter and SRAM shmoo plot are presented in Figs. 5.11 and 5.12. No distinguishable difference is found between the two variants. Several reasons are considered. Noise in one phase is canceled out by that in the opposite phase. Only a small number of gates are propagating signals at any given moment, while all the rest are static and stabilizing the substrate potential with their junction capacitance. At lower supply voltages, smaller substrate noise is expected, due to the impact ionization and capacitance coupling between the drain and the substrate. Simultaneous switching of large number of data at a local area, however, should be treated with care.

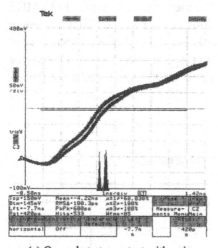
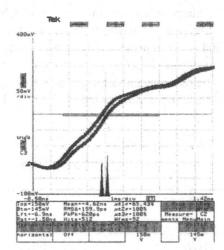

(a) One substrate contact with noise.   (b) 400 substrate contacts without noise.

Figure 5-11. Measured DLL tracking jitter.

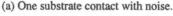
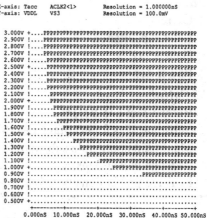

(a) One substrate contact with noise.   (b) 400 substrate contacts without noise.

Figure 5-12. Measured SRAM shmoo plot.

5.2.3.7 Technology Scaling of Reverse Body Bias

The power supply voltages will be lowered as the CMOS device is scaled. Is it then possible to make a large enough change in V_{TH} through the substrate bias even in low-voltage devices?

Suppose a scaling theory of VTCMOS based on the constant field MOS scaling theory where both device dimensions (which include channel length and width represented by x and oxide thickness t_{ox}) and device voltages (V) are scaled lineally by $1/\kappa$ (κ is a scaling factor larger than 1) and the substrate concentration density (N_A) is raised by κ. The resultant effect of the scaling is summarized in Table 5-1.

Table 5-1. VTCMOS scaling theory.

Parameter		Scaling Scenario
Device		
Horizontal size (L,W)	x	$1/\kappa$
Supply voltage	V	$1/\kappa$
Oxide thickness	t_{ox}	$1/\kappa$
Electric field across gate oxide	E	1
Substrate doping	N_A	κ
Circuit		
Current	$I \propto V^{1.4}/t_{ox}$	$1/\kappa^{0.4}$
Load capacitance	$C \propto x^2/t_{ox}$	$1/\kappa$
Gate delay	$t \propto CV/I$	$1/\kappa^{1.6}$
Power density	$p \propto VI/x^2$	$\kappa^{0.6}$
VTCMOS		
Body effect coefficient	$\gamma \propto t_{ox} N_A^{0.5}$	$1/\kappa^{0.5}$
Substrate bias for constant V_{TH} shift	$V_{BB} \propto 1/\gamma^2$	κ
Maximum applicable substrate bias	$V_{BB.max} \propto V$	$1/\kappa$
Leakage power density	$p_{leak} \propto V/x$	1
Leakage power density / Power density	p_{leak}/p	$1/\kappa^{0.6}$

Assumptions and reasons are as follows. (1) The transistor on-current is proportional to $(1/t_{ox})(W/L)(V_{DD}-V_{TH})^{1.4}$ [18]. (2) The body effect coefficient γ is given by Eq. (5.2). (3) V_{BB} required to raise V_{TH} by a certain amount is almost proportional to $1/\gamma^2$, which is derived from Eq. (5.2) by putting $\phi_b=0$.

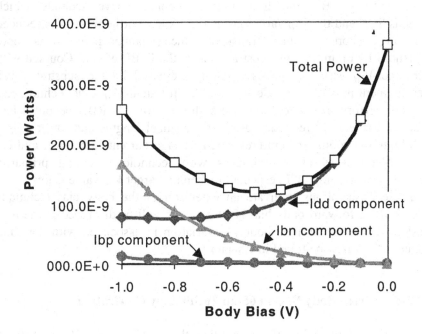

Figure 5-13. Optimum reverse body bias.

(4) The maximum V_{BB} caused by avalanche breakdown must be proportional to V_{DD}, because otherwise it will not be possible to perform a burn-in test. Table 5.1 shows that γ is reduced by $1/\sqrt{\kappa}$, and V_{BB} required to raise V_{TH} by a certain amount is increased by κ. On the other hand, an upper limit on V_{BB} is reduced by $1/\kappa$. In theory, it will eventually be difficult to cause large enough change in V_{TH} through RBB.

Furthermore, junction band-to-band tunneling leakage, as well as short channel effects (SCE) and drain-induced barrier lowering (DIBL), should be taken into consideration.

As V_{BB} increases, the sub-threshold leakage decreases, while the junction tunneling leakage increases. As a result, there is an optimum reverse body bias which minimizes the total chip leakage power, as depicted in Figure 5-13. Measurement shows [19]: (1) the junction leakage current is dominated by bulk band-to-band tunneling, not surface band-to-band tunneling that is called gate-induced drain leakage (GIDL), (2) the optimum RBB value reduces by about 2X per technology generation, and (3) the maximum achievable leakage power reduction by RBB diminishes by around 4X per generation under constant field technology scaling scenario. New junction engineering techniques to reduce the bulk band-to-band tunneling leakage current component across the junction are needed to preserve the effectiveness of RBB for standby leakage control.

In applying RBB, the drain-substrate depletion layer extends, which worsens SCE and the V_{TH} variations across a die. Furthermore, γ is reduced more in a shorter channel transistor, since channel potential is more influenced by drain than by substrate due to the DIBL effect. Coupled with SCE, the V_{TH} variation across a die is increased by the substrate bias. Measurement in 0.18μm single-V_{TH} and 0.13μm dual-V_{TH} logic technologies for high performance microprocessors shows [20]: (1) RBB becomes less effective for leakage reduction at shorter channel lengths and lower V_{TH} at both high and room temperatures when leakage currents are large, and (2) RBB effectiveness also diminishes with technology scaling primarily because of worsening SCE, especially when the target V_{TH} value is low.

These observations motivate an extension of the body bias technique from RBB to forward body bias (FBB), which will be discussed in the next session. More discussion about V_{TH} variation in associated with the short channel effect may be found in Chapter 6.

5.2.3.8 Reverse Body Bias in 65nm Technology Generation

The simplified scaling theory predicts that it will eventually be difficult to cause a large enough change in V_{TH} through RBB. In practice, RBB is still effective in the 65nm technology generation by careful channel engineering and V_{DD} control [21].

Measured I_{DS}-V_{GS} characteristics of an NMOS transistor whose channel length is 50nm and gate oxide thickness is 1.3nm is depicted in Figure 5-14. Different V_{DD}'s and V_{BB}'s are provided for different I_{on} and I_{off} targets in three operation modes; a high-speed mode, a nominal mode, and a power-save mode. In the nominal mode, the transistor is operated with V_{DD}=0.9V and V_{BB}=-0.5V for I_{on}=650μA and I_{off}=10μA, for standard performance. In the high-speed mode, the transistor is operated with V_{DD}=1.2V and V_{BB}=0V for I_{on}=1150μA and I_{off}=85μA, which is 75% higher drivability compared with the nominal mode. In the power-save mode the transistor is operated with V_{DD}=0.6V and V_{BB}=-2.0V for I_{on}=120μA and I_{off}=0.3μA, which is only 3% leakage current of that in the nominal mode. In this way, by varying V_{DD} and V_{BB}, transistor performance can be set for high performance and low power dissipation.

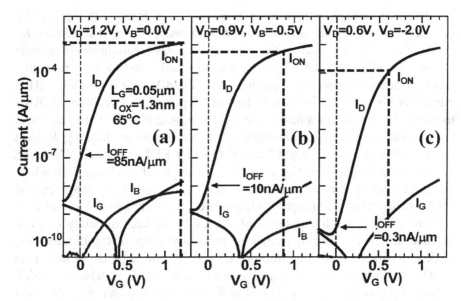

Figure 5-14. I_{DS}-V_{GS} characteristics of NMOS transistor of L=50nm and T_{ox}=1.3nm operated by different mode: (a) high-speed mode, (b) nominal model, and (c) power-save mode.

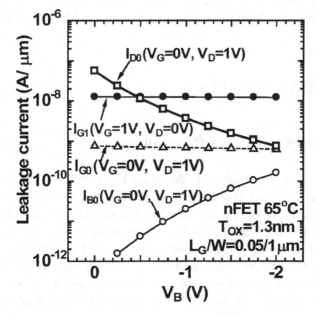

Figure 5-15. Leakage current dependence on V_{BB}: gate leakage I_{G0}, I_{G1}, sub-threshold leakage I_{D0}, and junction leakage I_{B0}.

Measured leakage current dependence on V_{BB} is depicted in Fig 5.15 by its four components; gate leakage current from and to the next-stage gate, I_{G0} (V_{GS}=0V and V_{DS}=1V) and I_{G1} (V_{GS}=1V and V_{DS}=0V), sub-threshold leakage current, I_{D0} (V_{GS}=0V and V_{DS}=1V), and the bulk band-to-band tunneling leakage current across the junction, I_{B0} (V_{GS}=0V and V_{DS}=1V). RBB efficiently reduces I_{D0} by two orders of magnitude. Even though RBB increases I_{B0}, it remains smaller than I_{D0} for V_{BB}<-2.0V. RBB has no effect on I_{G0} and I_{G1}, and I_{G1} dominates the leakage current, limiting leakage reduction effect by RBB when larger than -0.5V RBB is applied. However, I_{G1} reduces effectively as V_{DD} is lowered, as depicted in Figure 5-16 where measured leakage current dependence on V_{DD} is plotted. Leakage current can be reduced by two orders of magnitude by controlling both V_{DD} and V_{BB}.

Gate oxide reliability in terms of V_{TH} shifts are caused mainly by three mechanisms; 1) hot-carriers caused by lattice scattering in NMOS transistors, 2) time-dependent dielectric breakdown (TDDB) caused by lattice defects in PMOS transistors, and 3) negative bias temperature instability (NBTI) induced by positive charges created at Si/SiO_2 interface and in oxide under prolonged RBB in PMOS transistors. Measurement results show that RBB neither affects the NMOS hot-carrier lifetime nor the PMOS TDDB, but worsens the NBTI degradation. Lowering V_{DD} reduces NBTI effect significantly. Available V_{DD} and V_{BB} range for reliable transistor design with 1.1-1.6nm gate-SiON in association of hot carrier, TDDB, and NBTI is summarized in Figure 5-17. Reliability can be secured by controlling both V_{DD} and V_{BB}.

5.3 FORWARD BODY BIAS

5.3.1 Forward Body Bias Effectiveness

Extension of the drain-substrate depletion layer with reverse body bias (RBB) worsens the short channel effects (SCE), and hence, V_{TH} variation across a die. Furthermore, γ is lowered in a shorter channel transistor, since channel potential is influenced more by drain than substrate due to the drain-induced barrier lowering (DIBL) effect. SCE and DIBL are particularly severe in low-V_{TH} devices because reducing channel doping to lower V_{TH} causes the channel depletion depth to become larger. From these observations, the range of substrate biasing is extended from RBB to forward body bias (FBB). FBB is applied to a transistor with high V_{TH} to bring V_{TH} down to the target value. Experimental results by test chip measurement are summarized in this session. Details may be found in [22].

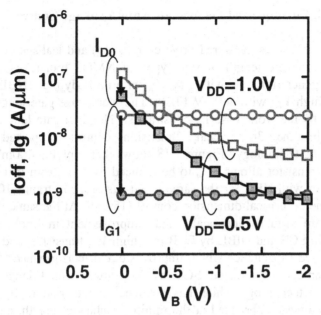

Figure 5-16. RBB at lower V_{DD} reduces leakage current by two orders of magnitude.

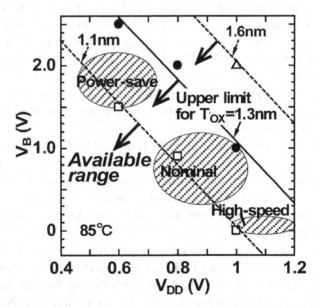

Figure 5-17. Available V_{BB} and V_{DD} range for reliable transistor design with 1.1-1.6nm gate-SiON in association of hot carrier, TDDB, and NBTI.

5.3.1.1 FBB for Improved Performance and Leakage Control

Figure 5-18 shows measured drive current (I_{on}) and leakage current (I_{off}) dependence on gate length of two types of NMOS transistors in 130nm technology generation, one with low V_{TH} with no body bias (NBB) and the other with high V_{TH} with 400mV FBB. The worst-case gate length, L_{wc}, is determined by the worst case I_{off} constraints. Nominal gate length, L_{nom}, is governed by the 3σ critical dimension control supported by the manufacturing technology. Figure 5-18 shows that applying 400mV FBB to a high-V_{TH} transistor allows L_{nom} to be reduced by 15% from that of a low-V_{TH} transistor with NBB for the same worst-case I_{off} constraint (100nA/μm) and the same 3σ critical dimension control (15nm). At the same time, their I_{on} values are virtually identical. This improvement in Lnom is due to reductions in SCE and DIBL by FBB to a high-V_{TH} transistor, and results in lower switched capacitance which improves circuit performance. However, taking advantage of reduced SCE by lowering channel length is only possible in a technology which has not reached the lithography limit. If devices are already fabricated at the minimum channel length, an alternate technique can be used in which FBB is used to match the I_{off} of a low-V_{TH} device with NBB at the same channel length. Figure 5-19 depicts measured I_{on} and I_{off} dependence on gate length of PMOS transistors, and shows that I_{on} of the FBB device is 7% larger than that of the low-V_{TH} device at identical L_{nom} (125nm) and the same worst-case I_{off} (400nA/μm). As a result, *CV/I* delay is 10% better for the FBB device. The FBB device, however, has larger junction capacitance ,due to reduced depletion widths around the source and drain junctions, and body effect. This reduces its *CV/I* delay advantage from 10% to 7% in inverter circuits, and to 5% in circuits containing transistor stacks.

The improvements in SCE and DIBL offered by FBB can be exploited to achieve a dual-V_{TH} technology. The same device is used in different parts of the design in either low-V_{TH} mode by applying FBB or in high-V_{TH} mode with NBB. Complexity of a dual-V_{TH} process is reduced since critical masking steps, needed for fabricating additional devices with different V_{TH} values, are eliminated.

Leakage reduction is enhanced by expanding the range of body biasing from RBB alone to RBB plus FBB. Leakage power during burn-in (110°C) and standby (27°C) can be reduced by withdrawing forward bias from the FBB device, and then applying RBB to both FBB and NBB devices. Leakage reduction of the FBB device from active to idle mode is 30X at 110°C for burn-in, and 20X at 27°C for standby, as depicted in Figure 5-20.

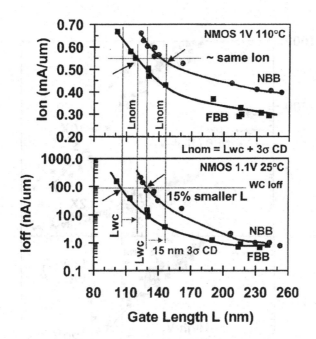

Figure 5-18. NMOS I_{on} and I_{off} versus L for 400mV FBB and NBB.

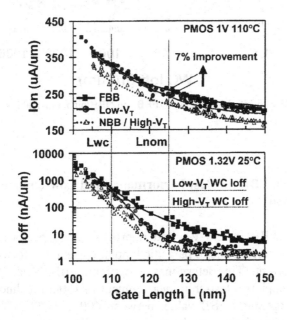

Figure 5-19. PMOS I_{on} and I_{off} versus L for 400mV FBB and NBB/high-V_{TH}, and low-V_{TH}.

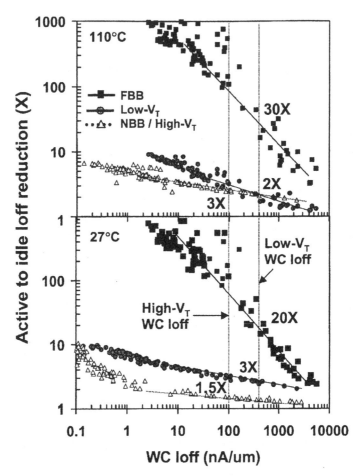

Figure 5-20. Active-to-idle I_{off} reduction of FBB and NBB/high-V_{TH}, and low-V_{TH} devices.

5.3.1.2 Optimum FBB for Best Performance and Low-temperature Operation

Even though FBB lowers V_{TH} and improves circuit performance, FBB increases leakage current due to parasitic bipolar current and forward source-body junction current. This determines an optimum FBB value.

Figure 5-21 depicts a test-chip measurement in 150nm technology. It is shown that the optimum FBB value, between 400 and 500 mV at 110°C, provides maximum frequency improvement (13%). The total switched

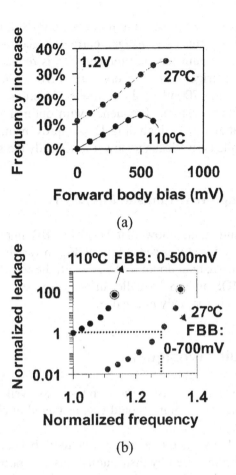

Figure 5-21. (a) Dependence on optimal FBB on maximum operating temperature. (b) Leakage versus frequency for FBB at 27°C and 110°C.

capacitance and switching energy are 10% higher because of larger junction capacitance, larger average gate capacitance at lower V_{TH}, and increased short-circuit current. Although active leakage power, including sub-threshold leakage, parasitic bipolar current and forward source-body junction current, increases by 10-100X, it remains sufficiently small compared to switching power. For bias values larger than this optimum, junction capacitance, body effect, and source-body junction forward current increases rapidly and fully negate any delay improvements induced by further V_{TH} reduction. Active leakage power also becomes an unacceptably large fraction of the total power. Therefore, for designs operating at a maximum junction temperature of 110°C, the desired FBB value is 450mV with ±50mV tolerance.

Lowering the operation temperature improves transistor drive current and circuit performance, reduces leakage currents, and improves device reliability. If the maximum junction temperature is reduced to 27°C using active cooling and refrigeration, the optimum FBB value increases from 450mV at 110°C to 700mV, since diode junction currents reduce exponentially with temperature. Frequency with optimum FBB at 27°C is 20% better than that at 110°C (including 15% improvements by temperature lowering alone), while active leakage power is virtually the same.

5.3.1.3 FBB for Improved Transistor Mismatch

Test chip measurements show that because FBB improves the device short-channel effects, it reduces sensitivity of V_{TH} to variation in gate length, oxide thickness, and channel doping. As a result, die-to-die V_{TH} variation is 36% smaller in PMOS and 48% smaller in NMOS when FBB is used, even with ±20% variation in the body bias value.

5.3.2 FBB Circuit Techniques

Sufficient robustness against various noises, as well as variations in process, voltage, and temperature should be considered in circuit design [23, 24].

Variation of the body-to-source voltage due to global coupling and power supply line noise is minimized by distributing bias generators, as depicted in Figure 5-22. A central bias generator uses a scaled bandgap circuit to generate a PVT-insensitive 450mV voltage with reference to V_{DDA}[25]. This reference voltage is routed to 24 local bias generators distributed around a digital core of a chip. Global routing of this 450mV differential reference voltage uses V_{DDA} tracks on both sides for proper shielding and adequate common-mode noise rejection. Each local bias generator has a reference translation circuit that converts the V_{DDA}-450mV reference to a voltage that is 450mV below the local V_{DD}. This voltage is driven by a buffer stage and routed locally to PMOS devices to provide 450mV FBB during active operation. Local body bias routing tracks are placed adjacent to the local V_{DD} track to improve common-mode noise rejection, and thus reduce noise-induced variations in the target 450mV V_{BB} in the biased PMOS devices. The voltage buffer and the local decoupling capacitor at the buffer output are designed to minimize V_{BB} variations induced by local coupling and V_{DD} noise, with a small area and power overhead. Full-chip area overhead of the biasing circuitry is 2% and power overhead is 1%.

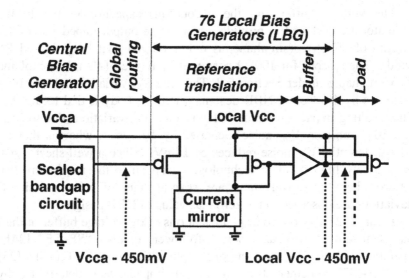

Figure 5-22. Bias generation and distribution.

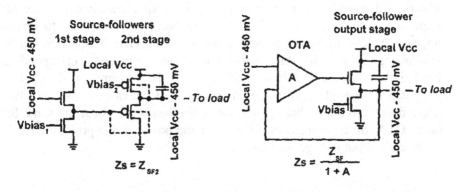

Figure 5-23. Bias generation and distribution.

The voltage buffer and the decoupling capacitor in the local bias generator are designed to provide a worst-case output impedance of 100kΩ per µm of effective (simultaneous unidirectional switching) biased PMOS width. The reasons for 100kΩ per µm design rule-of-thumb is explained in the next chapter under Section 6-3. The total V_{BB} variation induced by noise increases by 4% from a NBB design, where the body is tied locally to V_{DD}. The resulting impact on circuit delay is 1%. V_{BB} variations due to coupling and 10% common V_{DD} noise increases by 10-20mV, whereas that due to 10% differential V_{DD} noise reduces by 17mV. Since n-well sheet resistance is relatively high in logic technologies, and since the maximum distance allowed between n-well taps are several tens of microns, significant deviations are observed in the zero bias value of NBB designs.

Figure 5-23 shows two implementations of the voltage buffer in the local bias generators: two stage source-follower, namely "SF+SF" [24], and operational transconductance amplifier plus SF, namely "OTA+SF" [23]. In the "SF+SF" implementation, the overall impedance is determined by the output impedance of the second stage, while the first stage is designed to meet bandwidth requirements. FBB, already available in the local bias generators, is used for the PMOS devices in the output SF stage to improve gm by 30%, thus reducing the output impedance for the same area. The OTA+SF implementation uses a high-gain OTA and an output SF stage. The overall impedance is determined by the output impedance of the SF stage and the voltage gain of OTA. The design is optimized to obtain impedance less than the target value up to 10GHz frequency, while minimizing area. In this optimization, the gain and corresponding bandwidth of OTA are traded off against the amount of decoupling capacitance needed at the buffer output. Comparisons of the two implementations show that full-chip area overheads are about the same for both. OTA+SF consumes double the area and power, while providing better accuracy in input voltage tracking.

5.3.3   FBB Performance and Penalty

5.3.3.1   1GHz Communications Router Chip in 150nm CMOS

A 6.6M-transistor communications router chip, with on-chip circuitry to provide FBB during active operation and ZBB during standby mode, is implemented in a 150nm CMOS technology [23]. FBB is withdrawn during standby mode to reduce leakage power. Power and performance of the chip are compared with the original design that has no body bias (NBB). The FBB and NBB router chips reside adjacent to each other on the same reticle to allow accurate comparisons by measurement. Body bias is used for the PMOS devices in the digital core of the chip.

Maximum frequency (F_{max}) of the NBB and FBB router chips are compared from 0.9V to 1.8V V_{DD} at 60°C in Figure 5-24. The FBB chip with FBB achieves 1GHz operation at 1.1V, compared to 1.25V required for the NBB chip. As a result, switching power is 23% smaller at 1GHz. The frequency of the FBB chip is 33% higher than the NBB chip at 1.1V. The frequency improvement is more pronounced as V_{DD} is further reduced. Also, frequencies of the NBB chip and the body bias chip with ZBB are identical, indicating that there is no performance impact of potentially larger V_{DD} noise in the FBB design due to the absence of n-well to substrate junction capacitance for local V_{DD} decoupling, or of potentially larger V_{BS} noise due to separation of the well contacts Chip leakage currents are measured for 74 dies on a wafer with FBB and ZBB. Leakage current during active mode is set by FBB, which is withdrawn in standby mode to reduce leakage. 2x to 8x leakage reduction is achieved.

5.3.3.2 5GHz 32bit Integer Execution Core in 130nm CMOS

A 32bit integer execution core containing an ALU and a 32-entry x 32-bit register file (RF), with on-chip FBB circuitry, is implemented in a 130nm CMOS technology [24]. Body bias improvement measurements showing frequency vs. supply voltage measurements of ALU and RF are shown in Figure 5-25. Applying 450mV FBB to both NMOS and PMOS transistors allows 5GHz core frequency to be achieved at lower V_{DD} values for both ALU and RF. V_{DD} for 5GHz operation is reduced from 1.05V to 0.95V for ALU, a 9.5% reduction, and from 1.43V to 1.37V for RF, a 4.2% reduction. The power consumption of RF reduces by 6% by FBB.

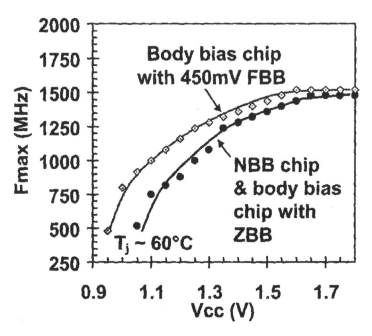

Figure 5-24. Frequency vs. V_{DD} of FBB and NBB chips.

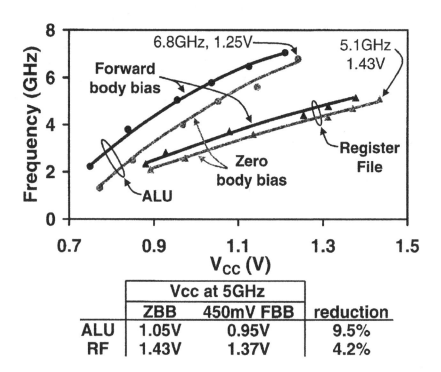

	Vcc at 5GHz		reduction
	ZBB	450mV FBB	
ALU	1.05V	0.95V	9.5%
RF	1.43V	1.37V	4.2%

Figure 5-25. ALU and register file frequency vs. V_{DD}.

5.4 FUTURE DIRECTIONS

Body biasing for reducing power dissipation of nanoscale CMOS logic chips has been described. It should be noted that body biasing is also applied to memory design [26]. Intensive discussion about low power memory design may be found in Chapter 7.

We have seen that the voltage range of body biasing has extended from RBB to FBB. The purpose of body biasing is also extending from reducing standby leakage current, to lowering maximum power dissipation by compensating ΔV_{TH} [27], and in future, for extending battery life by adjusting V_{TH} in accordance with work load [28].

Various adaptive control techniques are developed. Closed loop feedback control with a leakage monitor or speed monitor [29, 30] has been often used before, but open loop control with table lookup may be employed in future. The target of the monitor will probably extend from leakage current and circuit speed [29, 30], to workload [28], temperature, and reliability. Objectives of the control are extending to V_{DD} and frequency in addition to V_{TH}. In controlling both V_{DD} and V_{TH}, care should be paid to prevent oscillation between the two controls. Circuit speed is dependent on both V_{DD} and V_{TH}. Leakage current is also dependent on both of them due to DIBL in recent devices. The control method is also extending from analog to digital [27], and hardware to software [28]. Granularity of control in terms of space and time is becoming more fine; from chip level to block level [27, 31], and from microsecond to nanosecond ranges [31]. For instance, a self-adjusted FBB scheme in Figure 5-26 is employed for gated body, since the total current for generating FBB is limited by a current source in a controller such that the dc current does not dominate the total current dissipation, independent of the number of transistors in a block under the FBB control [31].

We discussed a technology in this chapter in which the body is biased uniformly all over the chip by monitoring chip leakage current and feedback control in analog domain. The future trend is to employ multiple biases by monitoring speed and PVT variations by digital control and software. For instance, a threshold voltage hopping (V_{TH}-hopping) is effective in low V_{DD} designs where V_{TH} is low and active leakage component is dominant in total power consumption [27]. V_{TH} is dynamically controlled through software, depending on a workload. As depicted in Figure 5-27, it can save 82% power dissipation compared with a fixed low-V_{TH} design in a 0.5V V_{DD} regime for multimedia applications. More discussion about adaptive control may be found in Chapter 6.

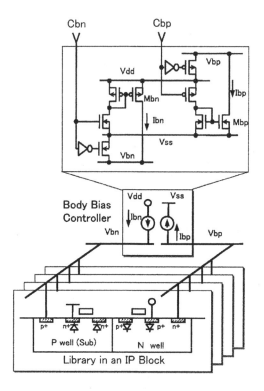

Figure 5-26. Self

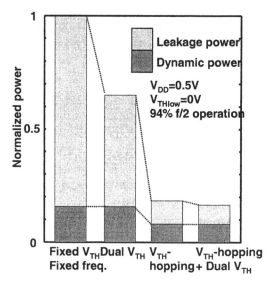

Figure 5-27. Power dissipation of designs with fixed V_{TH}, dual-V_{TH}, and V_{TH}-hopping.

REFERENCES

[1] T. Kuroda and T. Sakurai, "Overview of Low-Power ULSI Circuit Techniques," *IEICE Trans. Electron.*, vol. E78-C, no. 4, pp. 334-344, Apr. 1995.

[2] A. P. Chandrakasan and R. W. Brodersen, "Low Power Digital CMOS Design," *Kluwer Academic Publishers*, 1995.

[3] A. Chandrakasan, W. J. Bowhill, and F. Fox, *Design of High-Performance Microprocessor Circuits*, IEEE Press, Piscataway, NJ, 2001.

[4] M. Pedram and J. Rabaey, *Power Aware design Methodologies*, Kluwer Academic Publishers, 2002.

[5] P. Maxwell, R. Aitken, K. Kollitz, and A. Brown, "IDDQ and ac scan: the war against unmodelled defects," *Proc. Int. Test Conf.*, pp. 250-258, Nov. 1996.

[6] S-W. Sun and P. G. Y. Tsui, "Limitation of CMOS Supply-Voltage Scaling by MOSFET Threshold-Voltage Variation," *IEEE J. Solid-State Circuits*, vol. 30, no. 8, pp. 947-949, Aug. 1995.

[7] T. Kuroda and T. Sakurai, "Threshold-voltage control schemes through substrate-bias for low-power high-speed CMOS LSI design," *J. VLSI Signal Processing Systems, Kluwer Academic Publishers*, vol.13, no. 2/3, pp. 191-201, Aug./Sep. 1996.

[8] T. Kuroda, T. Fujita, S. Mita, T. Nagamatu, S. Yoshioka, K. Suzuki, F. Sano, M. Norishima, M. Murota, M. Kako, M. Kinugawa, M. Kakumu, and T. Sakurai, "A 0.9V 150MHz 10mW 4mm^2 2-D discrete cosine transform core processor with variable-threshold-voltage scheme," *IEEE J. Solid-State Circuits*, vol. 31, no. 11, pp. 1770-1779, Nov. 1996.

[9] T. Kobayashi and T. Sakurai, "Self-adjusting threshold-voltage scheme (SATS) for low-voltage high-speed operation," *Proc. CICC'94*, pp. 271-274, May 1994.

[10] K. Seta, H. Hara, T. Kuroda, M. Kakumu, and T. Sakurai, "50% active-power saving without speed degradation using standby power reduction (SPR) circuit," *ISSCC Dig. Tech. Papers*, pp. 318-319, Feb. 1995.

[11] T. Kuroda, T. Fujita, T. Nagamatu, S. Yoshioka, T. Sei, K. Matsuo, Y. Hamura, T. Mori, M. Murota, M. Kakumu, and T. Sakurai, "A high-speed low-power 0.3μm CMOS gate array with variable threshold voltage (VT) scheme," *Proc. CICC'96*, pp. 53-56, May 1996.

[12] T. Kuroda, T. Fujita, S. Mita, T. Mori, K. Matsuo, M. Kakumu, and T. Sakurai, "Substrate noise influence on circuit performance in variable threshold-voltage scheme," *Proc. ISLPED'96*, pp. 309-312, Aug. 1996.

[13] R. D. Pashley and G. A. McCormick, "A 70-ns 1K MOS RAM," *ISSCC Dig. Tech. Papers*, pp. 138-139, Feb. 1976.

[14] H. Makino, Y. Tsujihashi, K. Nii, C. Morishima, Y. Hayakawa, T. Shimizu, and T. Arakawa, "An auto-backgate-controlled MT-CMOS circuit," *Symposium on VLSI Technology Dig. Tech. Papers*, pp. 42-43, June 1998.

[15] M. Takahashi, M. Hamada, T. Nishikawa, H. Arakida, Y. Tsuboi, T. Fujita, F. Hatori, S. Mita, K. Suzuki, A. Chiba, T. Terasawa, F. Sano, Y. Watanabe, H. Momose, K. Usami, M. Igarashi, T. Ishikawa, M. Kanazawa, T. Kuroda, and T. Furuyama, "A 60mW MPEG4 video codec using clustered voltage scaling with variable supply-voltage scheme," *ISSCC Dig. Tech. Papers*, pp. 34-35, Feb. 1998.

[16] K. Kanda, K. Nose, H. Kawaguchi, and T. Sakurai, "Design Impact of Positive Temperature Dependence of Drain Current in Sub 1V CMOS VLSI's," *Proc. CICC'99*, pp. 563-566, May 1999.

[17] H. Kirsch, D. Clemons, S. Davar, J. Harman, C. Holder, W. Hunsicker, F. Procyk, J. Stefany, and D. Yaney, "A 1Mb CMOS DRAM," *ISSCC Dig. Tech. Papers*, pp. 256-257, Feb. 1985.

[18] T. Sakurai and A. R. Newton, "Alpha-power law MOSFET model and its applications to CMOS inverter delay and other formulas," *IEEE J. Solid-State Circuits*, vol. 25, no. 2, pp. 584-594, Apr. 1990.

[19] A. Keshavarzi, S. Narendra, S. Borkar, C. Hawkins, K. Royi, and V. De, "Technology scaling behavior of optimum reverse body bias for standby leakage power reduction in CMOS IC's," *Proc. Low Power Electronics and Design*, pp. 252-254, Aug. 1999.

[20] A. Keshavarzi, S. Ma, S. Narendra, B. Bloechel, K. Mistry, T. Ghani, S. Borkar, and V. De, "Effectiveness of reverse body bias for leakage control in scaled dual Vt CMOS ICs," *Proc. Low Power Electronics and Design*, pp. 207-212, Aug. 2001.

[21] M. Togo, T. Fukai, Y. Nakahara, S. Koyama, M. Makabe, E. Hasegawa, M. Nagase, T. Matsuda, K. Sakamoto, S. Fujiwara, Y. Goto, T. Yamamoto, T. Mogami, M. Ikeda, Y. Yamagata, and K. Imai, "Power-aware 65nm node CMOS technology using variable V_{DD} and back-bias control with reliability consideration for back-bias mode," *Symposium on VLSI Technology Dig. Tech. Papers*, pp. 88-89, June 2004.

[22] S. Narendra, A. Keshavarzi, B.A. Bloechel, S. Borkar, and V. De, "Forward body bias for microprocessors in 130-nm technology generation and beyond," *IEEE J. Solid-State Circuits*, vol. 38, no. 5, pp. 696-701, May 2003.

[23] S. Narendra, M. Haycock, V. Govindarajulu, V. Erraguntla, H. Wilson, S. Vangal, A. Pangal, E. Seligman, R. Nair, A. Keshavarzi, B. Bloechel, G. Dermer, R. Mooney, N. Borkar, S. Borkar, and V. De, "1.1 V 1 GHz communications router with on-chip body bias in 150 nm CMOS," *ISSCC Dig. Tech. Papers*, pp. 270-271, Feb. 2002.

[24] S. Vangal, M.A. Anders, N. Borkar, E. Seligman, V. Govindarajulu, V. Erraguntla, H. Wilson, A. Pangal, V. Veeramachaneni, J. Tschanz, Y. Ye, D. Somasekhar, B. Bloechel, G. Dermer, R.K. Krishnamurthy, K. Soumyanath, S. Mathew, S. Narendra, M. Stan, S. Thompson, V. De, and S. Borkar, "5-GHz 32-bit integer execution core in 130-nm dual-V_T CMOS," *IEEE J. Solid-State Circuits*, vol. 37, no. 11, pp. 1421-1432, Nov. 2002.

[25] H. Banba, H.Shiga, A. Umezawa, T. Miyabe, T. Tanzawa, S. Atsumi, and K. Sakui, "A CMOS band-gap reference circuit with sub-1V operation," *Symposium on VLSI Circuits Dig. Tech. Papers*, pp. 228-229, June 1998.

[26] H. Kawaguchi, Y. Itaka and T. Sakurai, "Dynamic Leakage Cut-off Scheme for Low-Voltage SRAM's," *Symposium on VLSI Circuits Dig. Tech. Papers*, pp.140-141, June 1998.

[27] J. Tschanz, J. Kao, S. Narendra, R. Nair, D. Antonladls, A. Chandrakasan, and V. De, "Adaptive body bias for reducing impacts of doe-to-deiand within-die parameter variations on microprocessor frequency and leakage," *IEEE J. Solid-State Circuits*, vol. 37, no. 11, pp. 1396-1402, Nov. 2002.

[28] K. Nose, M. Hirabayashi, H. Kawaguchi, S. Lee, and T. Sakurai, "V_{TH}-hopping scheme to reduce sub-threshold leakage for low-power processors," *IEEE J. Solid-State Circuits*, vol. 37, no. 3, pp. 413-419, Mar. 2002.

[29] M. Miyazaki, G. Ono, T. Hattori, K. Shiozawa, K. Uchiyama, and K. Ishibashi, "A 100-MIPS/W microprocessor using speed-adaptive threshold-voltage CMOS with forward bias," *ISSCC Dig. Tech. Papers*, pp. 420-421, Feb. 2000.

[30] G. Ono and M. Miyazaki, "Threshold-voltage balance for minimum supply operation," *Symposium on VLSI Circuits Dig. Tech. Papers*, pp. 206-209, June 2002.

[31] K. Ishibashi, T. Yamashita, Y. Arima, I. Minematsu, and T. Fujimoto, "A 9μW 50MHz 32b adder using a self-adjusted forward body bias in SoCs," *ISSCC Dig. Tech. Papers*, pp. Paper#6.8, Feb. 2003.

Chapter 6

PROCESS VARIATION AND ADAPTIVE DESIGN

Siva Narendra§, James Tschanz, James Kao¶, Shekar Borkar, Anantha Chandrakasan*, and Vivek De
§*Tyfone, Inc., USA, Intel Corp., USA, ¶Silicon Labs, USA, and *Massachusetts Institute of Technology, USA*

6.1 INTRODUCTION

With technology scaling as the transistor dimensions are reduced, manufacturing process variations and its related impact on design margins becomes substantial. Design margin is defined as the additional performance capability above required standard basic system parameters that may be specified by a system designer to compensate for uncertainties. This is a deterministic method that is often used by designers to deal with statistical uncertainties and minimize design time. With increase in transistor parameter variation the needed design margins at the circuit level for deterministic design is becoming too large. Therefore adaptive designs that do not need large design margins are essential. In order to make the analysis easier at the circuit level of hierarchy, in this chapter the overall impact of transistor parameter variation is viewed as change in the effective threshold voltage.

One of the critical sources of transistor parameter variation is related to short channel effect (SCE). As the channel length approaches the source-body and drain-body depletion widths, the charge in the channel due to these parasitic diodes becomes comparable to the depletion charge due to the MOSFET gate-body voltage [1], rendering the gate and body terminals to be less effective. As the band diagram in Figure 1-7 indicated in Chapter 1, the finite depletion widths of the parasitic diodes do not influence the energy barrier height to be overcome for inversion formation in a long channel device. However, as the channel length becomes shorter both channel length and drain voltage reduce this barrier height, as presented in Chapter 1.

This two-dimensional effect causes the barrier height to be modulated by channel length variation resulting in threshold voltage variation as discussed in Chapter 1. The amount of barrier height lowering, threshold voltage variation, and gate and body terminal's channel control loss will directly depend on the charge contribution percentage of the parasitic diodes to the total channel charge.

In Chapter 1, Figure 1-10 illustrated measurements of 3σ threshold voltage variations for three device lengths in a 0.18 μm technology confirming this behavior. It is essential to mention that in sub-micron technologies variation in several physical and process parameters lead to variation in the electrical behavior of the MOS device. The discussions in this chapter will address variation in the electrical behavior manifested as threshold voltage variation because of parameter variation. In addition, the threshold voltage variations addressed here are due to short channel effect in scaled MOS devices and not on threshold voltage variation due to random dopant fluctuation effect. Random dopant fluctuation effect is expected to be one of the significant sources of threshold voltage variation in devices of small area [2].

With supply and threshold voltage scaling, control of manufacturing process variation manifested as effective threshold voltage variation becomes essential for achieving high yields and limiting worst-case leakage. Figure 6-1 illustrates the three transistor threshold voltage variation categories which impact circuit design.

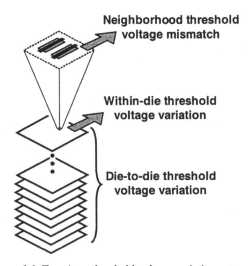

Figure 6-1. Transistor threshold voltage variation categories.

The die-to-die and within-die threshold voltage variation categories impact directly the variation in performance and leakage power of manufactured designs. Use of bi-directional adaptive forward and reverse body bias to limit the impact of threshold voltage variation is a more promising solution compared to the more traditional approach of using reverse body bias alone with process retargeting [3, 4].

Forward body bias is defined as a bias value applied to the body terminal of a transistor such that the body-to-source diode of the transistor is forward biased. Similarly reverse body bias is when this diode is reverse biased. Forward body bias can be used not only to reduce threshold voltage [5, 6], but also to reduce die-to-die and within-die threshold voltage variation. Bias circuit impedance requirements for on-chip body bias are also presented in this chapter. Use of both bi-directional adaptive body bias and adaptive supply voltage techniques provides the best adaptive design solution for temporally static process variation.

Temporally static manufacturing process variation is not the only source that impacts the circuit design margins. Time varying environmental parameters such as workload variation, temperature and supply voltage variations due to workload changes also impact the required design margins. While we do not cover adaptive solutions that deal with dynamic parameter variation, efficient and effective adaptive solutions that address them will become more essential in the future, and therefore this is an active area of research.

6.2 BI-DIRECTIONAL ADAPTIVE BODY BIAS

Both die-to-die and within-die V_t variations, which are becoming worse with technology scaling, impact clock frequency and leakage power distributions of microprocessors in volume manufacturing [7]. In particular, they limit the percentage of processors that satisfy both minimum frequency requirement and maximum active switching and leakage power constraints. Their impacts are more pronounced at the low supply voltages used in processors for mobile systems where the active power budget is limited by constraints imposed by heat removal, power delivery and battery life considerations.

In bi-directional adaptive body bias the mean V_t of all die samples are matched to the target V_t by applying both forward and reverse body bias. Forward body bias is applied to die samples that are slower than the target and reverse body bias is applied to die samples that are faster than the target, as shown in Figure 6-2. It is important to note that while forward bias reduces V_t it also increases the junction current. Hence, there is a maximum forward bias beyond which the junction current increase will inhibit proper operation of CMOS circuits. It has been determined that at a temperature of 110°C the maximum amount of forward bias that can be applied is 450 mV. This increases to 750 mV at an operating temperature of 30°C [8].

Since both V_t reduction and increase are possible, process re-targeting to reduce V_t is not required. By avoiding process re-targeting, increase in within-die V_t variation due increase in SCE for lower V_t transistors is prevented. In addition, the use of forward body bias reduces the diode depletion width and hence improves the SCE, reducing the within-die V_t variation. At the same time, the maximum reverse body bias required under bi-directional adaptive body bias is clearly smaller. So, this scheme will always scale better than the traditional adaptive body bias. This technique was first reported in [4] as a follow-up to [3]. In rest of this section, improvements over [4] will be presented.

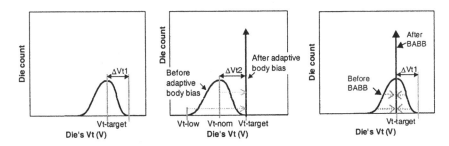

Figure 6-2. Die-to-die threshold voltage distributions (a) Conventional approach without adaptive body bias (b) traditional adaptive body bias approach – die sample that requires maximum reverse body bias is 2ΔVt2 away from Vt-target (c) bi-directional adaptive body bias approach – die sample that requires maximum reverse body bias is ΔVt1 away from Vt-target. Note: ΔVt2 > ΔVt1 since SCE of devices with lower V_t will be more.

A test chip (Figure 6-3) was implemented in a 150 nm CMOS technology to evaluate effectiveness of the bi-directional adaptive body bias technique for minimizing impacts of both die-to-die and within-die V_t variations on processor frequency and active leakage power [9]. The test chip contains 21 "sub-sites" distributed over a 4.5 x 6.7 mm^2 area in two orthogonal orientations.

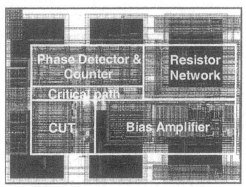

Figure 6-3. Chip micrograph of a sub-site.

Each sub-site has (i) a circuit block (CUT) containing key circuit elements of a microprocessor critical path, (ii) a replica of the critical path whose delay is compared against an externally applied target clock frequency (ϕ) by a phase detector, (iii) a counter which updates a 5-bit digital code based on the phase detector output, and (iv) a "resistor-ladder D/A converter + op-amp driver" which, based on the digital code, provides one of 32 different body bias values to PMOS transistors in both the CUT and the critical path delay element.

The circuit block diagram of each sub-site is shown in Figure 6-4. N-well resistors are used for the D/A converter implementation. For a specific externally applied NMOS body bias, this on-chip circuitry automatically generates the PMOS body bias that minimizes leakage power of the CUT while meeting a target clock frequency, as demonstrated by measurements in Figure 6-1. Different ranges of unidirectional – forward (FBB) or reverse (RBB) – or bi-directional body bias values (Figure 6-5) can be selected by using appropriate values of V_{REF} and V_{CCA}, and by setting a counter control bit.

Adaptive body biasing can also be accomplished by using the phase detector output (PD) to continually adjust off-chip bias generators through software control, instead of using the on-chip circuitry, until the frequency target is met.

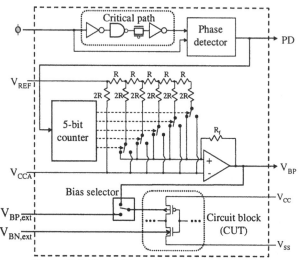

Figure 6-4. Circuit block diagram of each sub-site.

Clock frequency, switching power and active leakage power of the 21 CUT's per die are measured independently at 0.9V VCC and 110C, for 62 dies on a wafer. Die clock frequency is the minimum of the CUT frequencies, and active leakage power is sum of the CUT leakages. When no body bias (NBB) is used, 50% of the dies meet both the minimum frequency requirement and the maximum active leakage constraint set by a total power density limit of 20 W/cm^2 (Figure 6-6). Using 0.2V forward body bias (FBB) allows all of the dies to meet the minimum frequency requirement, but most of them fail to satisfy the leakage constraint. As a result, only 20% of the dies are acceptable even though variations are reduced slightly by FBB due to improved short-channel effects [4].

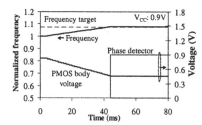

Bias Mode	Condition	Range
NBB → FBB	$V_{CCA} = V_{CC}$ $V_{REF} > V_{CCA}$	FBB: $0 \to V_{REF}\text{-}V_{CCA}$
NBB → RBB	$V_{CCA} = V_{CC}$ $V_{REF} < V_{CCA}$	RBB: $0 \to V_{CCA}\text{-}V_{REF}$
FBB → RBB	$V_{CCA} < V_{CC}$ $V_{REF} < V_{CCA}$	FBB: $V_{CC}\text{-}V_{CCA} \to$ RBB: $2V_{CCA}\text{-}V_{REF}\text{-}V_{CC}$

Figure 6-5. Demonstration of frequency adapting to meet target and list of possible on-chip bias modes.

Bi-directional ABB is used for both NMOS and PMOS devices to increase the percentage of dies that meet both frequency requirement and leakage constraint. For each die, we use a single combination of NMOS and

PMOS body bias values that maximize clock frequency without violating the active leakage power limit. As a result, die-to-die frequency variations (σ/μ) reduce by an order of magnitude, and 100% of the dies become acceptable (Figure 6-6). In addition, 30% of the dies are now in the highest frequency bin allowed by the power density limit when leakage is negligible.

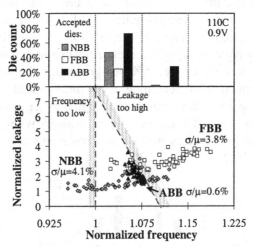

Figure 6-6. Die-to-die variation in frequency and leakage for no body bias (NBB), 0.2 V static forward body bias (FBB), and adaptive body bias applied to compensate die-to-die variation (ABB).

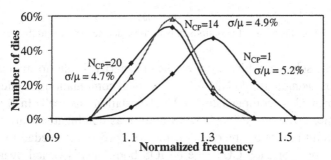

Figure 6-7. Frequency vs. number of critical paths that determine the frequency.

In a simpler ABB scheme, within-die variations can be neglected [4] and the required body bias for a die can be determined from measurements on a single CUT per die. However, test chip measurements in Figure 6-7 show that as the number of critical paths (NCP) on a die increases, WID delay variations among critical paths cause both μ and σ of the die frequency distribution to become smaller. This is consistent with statistical simulation results [7] indicating that the impact of WID parameter variations on die frequency distribution is significant. As NCP exceeds 14, there is no change in the frequency distribution with NCP. Therefore, using measurements of 21 critical paths on the test chip to determine die frequency is sufficiently accurate for obtaining frequency distributions of microprocessors, which contain 100's of critical paths.

Previous measurements [4] on 49-stage ring oscillators showed that σ of the WID frequency distribution is 4X smaller than σ of the device saturation current (ION) distribution. However, measurements on the test chip containing 16-stage critical paths (Figure 6-8) show that σ's of WID critical path delay distributions and NMOS/PMOS ION distributions are comparable. Since typical microprocessor critical paths contain 10-15 stages, and this number is reducing by 25% per generation, impact of within-die variations on frequency is becoming more pronounced. This is further evidenced by the fact that the number of acceptable dies reduces from 100% to 50% in the simpler ABB scheme which neglects within-die variations, although die count in the highest frequency bin increases from 0% to 11% when compared with NBB.

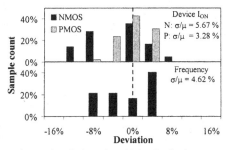

Figure 6-8. Comparison of variations in within-die device current and frequency.

The ABB scheme, which compensates primarily for die-to-die parameter variations by using a single NMOS/PMOS bias combination per die, can be further improved to compensate for WID variations as well. In this WID-ABB scheme, different NMOS/PMOS body bias combinations are used for different circuit blocks on the die. A triple-well process is needed for NMOS implementation. For each CUT, the NMOS body bias is varied over a wide range using an off-chip bias generator.

For each NMOS bias, the on-chip circuitry determines the PMOS bias that minimizes leakage power of the CUT while meeting a particular target frequency. The optimal NMOS/PMOS bias for the CUT at a specific clock frequency is then selected from these different bias combinations as the one that minimizes CUT leakage. This produces a distribution of optimal NMOS/PMOS body bias combinations for the CUT's on a die at a specific clock frequency. If the die leakage power exceeds the limit at that frequency, the target frequency is reduced and the process is repeated until we find the maximum frequency where the leakage constraint is also met.

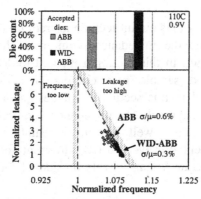

Figure 6-9. Die-to-die variation in frequency and leakage for adaptive body bias applied to (i) compensate die-to-die variation (ABB) and (ii) compensate within-die variation (WID-ABB).

WID-ABB reduces σ of the die frequency distribution by 50%, compared to ABB (Figure 6-9). In addition, virtually 100% of the dies are accepted in the highest possible frequency bin, compared to 30% for ABB. Distribution of optimal NMOS/PMOS body bias combinations (Figure 6-10) for a sample die in the WID-ABB scheme reveals that while RBB is needed for both PMOS and NMOS devices, FBB is used mainly for the PMOS devices. In addition, body bias values in the range of 0.5V RBB to 0.5V FBB are adequate. Finally, measurements (Figure 6-10) show that ABB and WID-ABB schemes need at least 300 mV and 100 mV body bias resolutions, respectively, to be effective. The 32 mV bias resolution provided by the on-chip circuitry in the test chip is, therefore, sufficient for both ABB and WID-ABB.

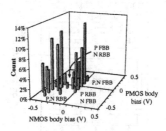

Bias	Die-to-die ABB		Within-die ABB	
resolution	dies, F > 1	σ/μ	dies, F > 1.075	σ/μ
0.5	79 %	2.87 %	2 %	1.89 %
0.3	100 %	1.47 %	66 %	0.50 %
0.1	100 %	0.58 %	97 %	0.25 %

Figure 6-10. Histogram of bias voltages within a die sample and effect of bias resolution on frequency distribution.

6.3 BODY BIAS CIRCUIT IMPEDANCE

Since adaptive body bias circuit techniques require on-chip biasing, it is important to determine the impedance requirement for the on-chip bias voltage generator circuit. In this section a method to determine proper bias circuit impedance and sample bias circuits are described. While the methodology described in this section is for design of forward body bias circuitry, the overall approach for impedance requirement will hold for reverse body bias circuits as well. The implementation to get the required impendence requirement however will be different for reverse body bias since it needs to generate voltage levels outside of the power supply rail levels.

To verify the design of the bias circuits a 6.6 million transistors communications router chip [10, 11, 12], with on-chip circuitry to provide forward body bias (FBB) [13] during active operation and zero body bias (ZBB) during standby mode, has been implemented in a 150 nm CMOS technology (Figure 6-11). FBB is applied during active mode and it is withdrawn during standby mode to reduce leakage power. Power and performance of the chip are compared with the original design that has no body bias (NBB). The FBB and NBB router chips reside adjacent to each other on the same reticle to allow accurate comparisons by measurement.

If the on-chip bias circuit has proper impedance then (i) FBB chip in FBB mode should increase the frequency of operation at a given supply voltage (V_{cc}) (ii) FBB chip in ZBB mode should have lower standby leakage and (iii) FBB chip with ZBB should have the same frequency of operation as that of the NBB chip on the same reticle.

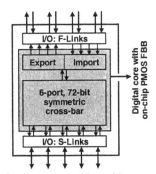

Figure 6-11. Communications router chip architecture with PMOS body bias.

In the FBB test chip, body bias is used for the PMOS devices in the digital core of the chip. Total biased PMOS transistor width is 2.2 meters. Body bias generator circuits and bias distribution across the chip have been optimized to minimize area overhead, and to provide a constant 450 mV FBB with sufficient robustness against various noises, as well as variations in process, V_{cc}, and temperature (PVT).

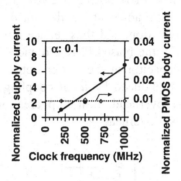

Figure 6-12. Measurement of body and Vcc current.

Test chip measurements (Figure 6-12) show that current in the body grid is at least two orders of magnitude smaller than the V_{cc} current across a range of operating frequencies. Therefore, overhead of body bias routing is minimal compared to the V_{cc} grid. Distributed bias generator architecture has been implemented to minimize variation of the body-to-source voltage (V_{bs}) due to *global* coupling and V_{cc} noises (Figure 6-13). A central bias generator (CBG) uses a scaled bandgap circuit [14] to generate a PVT-insensitive 450 mV voltage with reference to V_{cca}. This reference voltage is routed to 24 local bias generators (LBG), distributed around the digital core of the chip. Global routing of this 450 mV differential reference voltage uses V_{cca} tracks on both sides for proper shielding and adequate common-mode noise rejection.

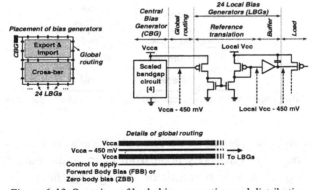

Figure 6-13. Overview of body bias generation and distribution.

Each LBG has a reference translation circuit that converts the V_{cca}-450 mV reference voltage to a voltage 450 mV below the local Vcc. This voltage is driven by a buffer stage and routed locally to the PMOS devices in the core to provide 450 mV FBB during active operation. Local body bias routing tracks are placed adjacent to the local V_{cc} tracks to improve common-mode noise rejection, and thus reduce noise-induced variations in the target 450 mV V_{bs} in the biased PMOS devices. The voltage buffer and the local decoupling capacitor at the buffer output have been designed to minimize V_{bs} variations induced by *local* coupling and V_{cc} noises, with a small area and power overhead. Full-chip area overhead of the biasing circuitry is 2% and power overhead is 1%.

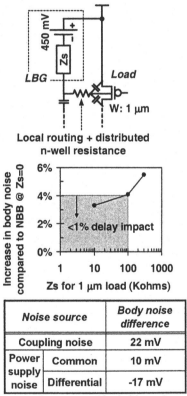

Figure 6-14. Buffer impedance requirements and body bias noise comparisons with NBB.

Noise source		Body noise difference
Coupling noise		22 mV
Power supply noise	Common	10 mV
	Differential	-17 mV

Three different sources of noise can induce variations in the target V_{bs} value. First, coupling to the body node from logic circuit output transitions can change V_{bs} of a victim transistor during switching. This noise is transmitted to the victim through the bias grid and the n-well. Circuit simulations in a 150 nm technology, with a two-dimensional distributed RC model for the n-well, show that the width of this noise pulse is several hundred pico-seconds for 1.5-2 KΩ/sq n-well sheet resistance. Therefore, this noise impacts switching delay of the victim circuit. However, since different circuits switch in opposite directions at the same time in a large logic design, a small fraction (<10%) of the total transistor width accounts for the simultaneous unidirectional switching that couples noise to the body. Second, V_{cc} noise common to both the V_{bs} generator and a biased logic circuit can causes V_{bs} to vary (common V_{cc} noise). Finally, any difference in V_{cc} values of the V_{bs} generator and the biased logic circuit induces variation in V_{bs} (differential V_{cc} noise).

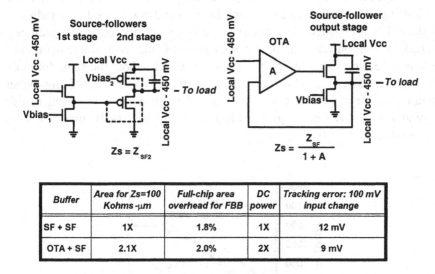

Buffer	Area for Zs=100 Kohms -μm	Full-chip area overhead for FBB	DC power	Tracking error: 100 mV input change
SF + SF	1X	1.8%	1X	12 mV
OTA + SF	2.1X	2.0%	2X	9 mV

Figure 6-15. LBG buffer implementations and comparisons.

The voltage buffer and the decoupling capacitor in the LBG have been designed to provide worst-case output impedance (Z_s) of 100 KΩ per μm of *effective* (simultaneous unidirectional switching) biased PMOS width. Simulations (Figure 6-14) for this design show that the total V_{bs} variation induced by all three noises increases by 4% from the NBB design, where the body is tied locally to V_{cc}. The resulting impact on circuit delay is 1%. V_{bs} variations due to coupling and 10% common V_{cc} noise increase by 10-20 mV, whereas that due to 10% differential V_{cc} noise reduces by 17 mV. Since n-well sheet resistance is relatively high in logic technologies, and since the maximum distance allowed between n-well taps are several tens of microns, significant deviations are observed (by simulations) in the zero bias value for NBB designs.

Figure 6-15 shows implementations of two different LBG buffers – the "SF+SF" on a 5GHz 32-bit integer execution core chip [15] in a 130nm dual-Vt CMOS technology, and the "OTA+SF" on this chip. In the SF+SF implementation, the overall impedance is determined by the output impedance Z_{SF2} (~$1/g_m$) of the second stage, while the first stage is designed to meet bandwidth requirements. FBB, already available in the LBG, is used for the PMOS devices in the output source-follower (SF) stage to improve g_m by 30%, thus reducing Zs for the same area.

The OTA+SF implementation uses a high-gain OTA and an output SF stage. The overall impedance is determined by the output impedance (Z_{SF}) of the SF stage and the voltage gain (A) of the OTA. The design is optimized to obtain impedance less than the target value up to 10 GHz frequency, while minimizing area. In this optimization, the gain and corresponding bandwidth of the OTA are traded-off against the amount of decoupling capacitance needed at the buffer output. Comparisons of the two implementations (Figure 6-15) show that full-chip area overheads are about the same for both. OTA+SF consumes double the area and power, while providing better accuracy in input voltage tracking. The OTA+SF design was used in this communications router chip as well as the adaptive body bias test chip described in Section 6.2.

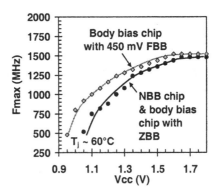

Figure 6-16. Frequency vs. Vcc of FBB and NBB chips.

Maximum frequency (F_{max}) of the NBB and FBB router chips are compared from 0.9 V to 1.8 V V_{cc} at 60°C (Figure 6-16). F_{max} values are measured by sending data in through the F-link input port and verifying data at the F-link output port after the data has passed through the import, crossbar and export units. The FBB chip with forward body bias achieves 1 GHz operation at 1.1 V, compared to 1.25 V required for the NBB chip and FBB chip with ZBB. As a result, switching power is 23% smaller at 1 GHz.

The frequency of the FBB chip is 33% higher than the NBB chip at 1.1 V. The frequency improvement is more pronounced as V_{cc} is further reduced. Also, there is no observable performance impact of potentially larger V_{cc} noise in the FBB design due to the absence of n-well to substrate junction capacitance for local Vcc decoupling.

Chip leakage currents are measured for 74 dies on a wafer with FBB and ZBB. Leakage current during active mode is set by FBB, which is withdrawn in standby mode to reduce leakage. Histogram (Figure 6-17) of active-to-standby leakage ratio – I(FBB)/I(ZBB) – shows 2X to 8X leakage reduction, with an average reduction of 3.5X. Clearly, this leakage reduction capability will not available in the NBB chip, if the performance is improved by lowering V_t in the process technology. The die micrograph and chip characteristics are shown in Figure 6-18.

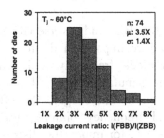

Figure 6-17. Leakage reduction from active to standby mode in FBB chips.

It was shown that (i) the FBB chip in FBB mode operates at 23% higher frequency at 1.1 V, (ii) the FBB chip in ZBB mode on an average has 3.5X lower standby leakage and (iii) the FBB chip with ZBB has virtually the same frequency of operation as that of the NBB chip at any given V_{cc}. This proves for the chosen load the method of targeting the bias circuit impedance to be 100 KΩ per μm of effective (simultaneous unidirectional switching) biased MOS width is a successful rule-of-thumb. One can arrive at similar guidelines depending on the noise and bias accuracy requirements.

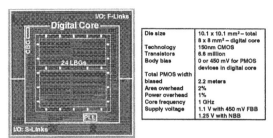

CBG – Central bias generator (836 μm x 267 μm)
LBGs – Local bias generators (156 μm x 68 μm each)

Figure 6-18. Micrograph of communications router chip with PMOS body bias and of chip characteristics.

6.4 ADAPTIVE SUPPLY VOLTAGE AND ADAPTIVE BODY BIAS

Bidirectional – forward (FBB) and reverse (RBB) – adaptive body bias (V_{bs}) has been used to reduce impacts of die-to-die and within-die (WID) parameter variations on clock frequency and active leakage power of microprocessors in volume manufacturing [4, 9] as was described in detail in Section 6.2. In this section, we investigate the effectiveness of adaptive supply voltage (V_{cc}) and frequency binning, used individually and in conjunction with adaptive V_{bs}, for improving distributions of die frequency and power in low power and high performance microprocessors. We compare usefulness of these schemes for maximizing the percentage of dies accepted in the highest frequency bin, subject to constraints of total active power, burn-in leakage power and standby leakage power.

The same test chip described in Section 6.2 is used to study the impact of using adaptive supply voltage along with adaptive body bias. Separate pads are available for V_{cc}, ground and body bias of NMOS and PMOS devices. Thus, a range of V_{cc} and V_{BS} values can be applied externally to each circuit under test (CUT), and their switching and leakage powers measured accurately. Frequency, power, and switched capacitance of the 21 CUT's per die are measured independently at 0.8V-1.6V V_{cc} and 40°C-110°C, with V_{BS} values ranging from 500mV RBB to 500mV FBB, for 62 dies on a wafer. Die clock frequency is the minimum of the CUT frequencies and die power is the sum of CUT powers.

6.4.1 Effectiveness of Adaptive Supply Voltage

Distributions of frequency, active power, switched capacitance and standby leakage power are shown in Figure 6-19 for a fixed 1.05 V V_{cc} and 0 V V_{BS}, chosen to maximize frequency of the median die for a power density limit of 10 W/cm², typical of low power microprocessors in mobile systems. We see that dies with higher frequencies have larger leakage and smaller switched capacitance.

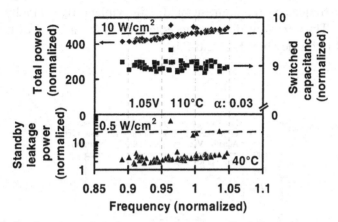

Figure 6-19. Total power, switched capacitance, and standby leakage power vs. frequency. V_{CC}=1.05V for maximum frequency of median die at 10 W/cm² total power limit.

Excessive leakage causes many dies to violate constraints of active power and standby leakage power (0.5 W/cm²). Many other dies satisfy the power constraints, but are significantly slower than the median die.

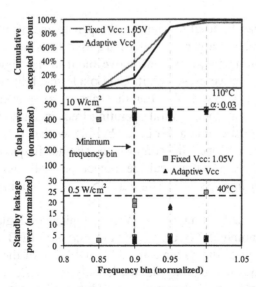

Figure 6-20. Cumulative accepted die count, total power, and standby leakage power vs. frequency bin for fixed V_{cc} and adaptive V_{cc}. V_{cc}=1.05 V for maximum frequency of median die at 10 W/cm² total power limit.

Since lowering frequency reduces switching power, active power limit can be satisfied for all dies by simply moving them to a frequency bin, equal to or below their natural operating frequencies, where the total power is less than the maximum allowed. However, some dies then fail to meet the minimum frequency requirement and still violate the standby leakage constraint. Thus, 95% of dies are accepted. In addition, 37% of dies are in the lowest frequency bin (Figure 6-20).

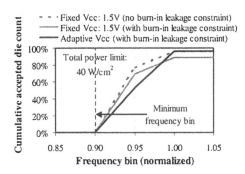

Figure 6-21. Cumulative accepted die count vs. frequency bin with fixed V_{CC} and adaptive V_{CC}. Fixed V_{CC} is 1.5V for maximum frequency of median die at 40 W/cm^2 total power limit.

Adaptive V_{cc} can be used to improve the percentage of dies accepted in higher frequency bins. Larger V_{cc} values are used for slow dies to increase their natural operating frequency and move them to the highest frequency bin allowed by the active power limit. Gate oxide reliability considerations limit the maximum allowed V_{cc}. For dies above the power limit, V_{cc} is reduced in tandem with their natural operating frequencies so that they meet the active power constraint at a frequency bin higher than that achievable by frequency reduction alone. In contrast to simple frequency reduction, lowering V_{cc} reduces standby leakage power as well. As a result, the accepted die count improves to 98%, with 15% in the minimum frequency bin.

For a power density limit of 40 W/cm^2, typical of high performance microprocessors, nominal V_{cc} of 1.5 V and 100 mV FBB are chosen to maximize frequency of the median die. Although 97% of dies are accepted when simple frequency binning is used, only 19% are in the highest frequency bin (Figure 6-21). Imposing an additional burn-in leakage power limit of 5 W/cm^2 reduces the percentage of accepted dies to 89%. Adaptive V_{cc} improves the accepted die count to 97%, with 44% of dies in the highest frequency bin. However, effectiveness of adaptive V_{cc} depends critically on the V_{cc} resolution.

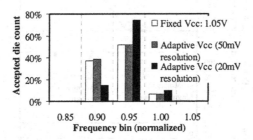

Figure 6-22. Accepted die count vs. frequency bin for 1.05V fixed V_{cc}, and adaptive V_{cc} with 20 mV and 50 mV resolution.

Using 50 mV V_{cc} resolution, instead of 20 mV, renders this technique ineffective (Figure 6-22) for compensating across-wafer variations, but may be adequate when larger variations, across multiple wafers and lots, are considered.

6.4.2 Adaptive Supply Voltage and Adaptive Body Bias

Using adaptive V_{cc} in conjunction with adaptive V_{bs} (adaptive $V_{cc}+V_{bs}$) is more effective than using either of them individually (Figure 6-23). In this combined scheme, a single V_{cc} and NMOS/PMOS V_{bs} combination is used per die to move it to the highest frequency bin subject to the active power limit. Adaptive V_{bs} uses FBB to speed up dies that are too slow, and RBB to reduce frequency and leakage power of dies that are too fast and leaky.

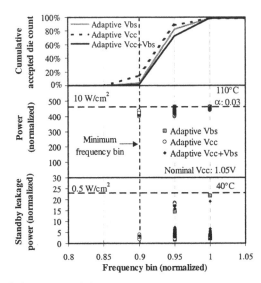

Figure 6-23. Cumulative accepted die count, total power, and standby leakage power vs. frequency bin for adaptive V_{cc}, adaptive V_{bs}, and adaptive $V_{cc}+V_{bs}$.

Adaptive $V_{cc}+V_{bs}$, on the other hand, recovers these dies above the active power limit by (1) first lowering V_{cc} and natural operating frequency together to bring the sum total of their switching and leakage powers well below the active power limit, and (2) then applying FBB to speed them up, and move them to the highest frequency bin allowed by the active power limit. As a result, more dies use lower V_{cc} values than adaptive V_{cc} (Figure 6-24). In addition, more dies use FBB, instead of RBB, compared to adaptive V_{bs} (Figure 6-25).

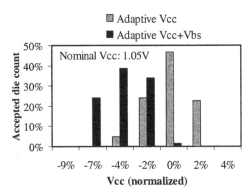

Figure 6-24. Distribution of V_{cc} values for adaptive V_{cc} and adaptive $V_{cc}+V_{bs}$.

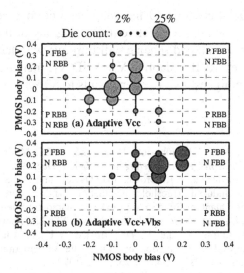

Figure 6-25. Distribution of optimal NMOS and PMOS body bias values for **(a)** adaptive V_{cc} and **(b)** adaptive $V_{cc}+V_{bs}$.

Since effectiveness of RBB for leakage power reduction diminishes with technology scaling [3], adaptive $V_{cc}+V_{bs}$ will be more effective in future technology generations. Combining adaptive V_{cc} with adaptive WID-V_{bs} can compensate for within-die variations as well. In this adaptive $V_{cc}+$WID-V_{bs} scheme, different NMOS/PMOS body bias combinations are used for different circuit blocks on a die, while a single V_{cc} is used for all circuit blocks. A triple-well process is needed for NMOS implementation. Using this technique increases the number of dies accepted in the highest two frequency bins from 26% to 80% (Figure 6-26).

Number of dies in F=1 and F=1.05 frequency bins	
6%	Fixed V_{DD}
10%	Adaptive V_{DD}
16%	Adaptive V_{BS}
26%	Adaptive $V_{DD} + V_{BS}$
80%	Adaptive V_{DD} + Within-die V_{BS}

Figure 6-26. Comparison of different schemes in its ability to increase number of dies in the highest frequency bins by reducing impact of variation. This comparison is in 150 nm CMOS with 10 W/cm^2 active & 0.5 W/cm^2 standby power density limits, with 100 mV body bias and 20 mV supply voltage resolutions.

Adaptive techniques are useful for reducing impacts of die-to-die and WID parameter variations on frequency, active power and leakage power distributions of both low power and high performance microprocessors that use nanoscale MOS transistors. Using adaptive Vcc together with adaptive V_{bs} or WID-V_{bs} is much more effective than using any of them individually. Effectiveness of the combined scheme will improve with technology scaling since fewer dies need RBB. Adaptive V_{cc}+WID-V_{bs} increases the number of dies accepted in the highest two frequency bins to 80%.

REFERENCES

[1]   H.C. Poon, L.D. Yau, R.L. Johnston, D. Beecham, "DC Model for Short-Channel IGFET's," *Intl. Electron Devices Meeting*, pp. 156-159, Dec. 1973.

[2]   A. Asenov, G. Slavcheva, A.R. Brown, J.H. Davies, and S. Saini, "Increase in the Random Dopant Induced Threshold Fluctuations and Lowering in Sub-100 nm MOSFETs due to Quantum Effects: A 3-D Density-Gradient Simulation Study," *IEEE Transactions on Electron Devices*, vol. 48, no. 4, pp. 722-729, April 2001.

[3]   S. Narendra, D. Antoniadis, and V. De, "Impact of Using Adaptive Body Bias to Compensate Die-to-die Vt variation on Within-die Vt variation," *Intl. Symp. Low Power Electronics and Design*, pp. 229-232, Aug. 1999.

[4]   M. Miyazaki, G. Ono, T. Hattori, K. Shiozawa, K. Uchiyama, and K. Ishibashi, "A 1000-MIPS/W Microprocessor using Speed Adaptive Threshold-Voltage CMOS with Forward Bias," *Intl. Solid-State Circuits Conf.*, pp. 420-421, 2000.

[5]   V. De, "Forward Biased MOS Circuits," *United States Patent, Patent number: 6,166,584*, Filed: June 1997, Issued: Dec. 2000.

[6]   C. Wann, J. Harrington, R. Mih, S. Biesemans, K. Han, R. Dennard, O. Prigge, C. Lin, R. Mahnkopf, and, B. Chen, "CMOS with Active Well Bias for Low-Power and RF/Analog Applications," *Symp. on VLSI Technology*, pp. 158-159, 2000.

[7]   K. Bowman, S. Duvall, and J. Meindl, "Impact of die-to-die and within-die parameter fluctuations on the maximum clock frequency distribution", *Intl. Solid-State Circuits Conf.*, pp. 278-279, 2001.

[8]   A. Keshavarzi, S. Narendra, B. Bloechel, S. Borkar, and V. De, "Forward Body Bias for Microprocessors in 130nm Technology Generation and Beyond," *Symp. on VLSI Circuits*, 2002.

[9]   J. Tschanz, J. Kao, S. Narendra, R. Nair, D. Antoniadis, A. Chandrakasan, and V. De, "Adaptive Body Bias for Reducing Impacts of Die-to-Die and Within-Die Parameter Variations on Microprocessor Frequency and Leakage," *Intl. Solid-State Circuits Conf.*, 2002.

[10] S. Narendra et.al., "1.1V 1GHz Communications Router with On-Chip Body B\ias in 150nm CMOS," *Intl. Solid-State Circuits Conf.*, 2002.

[11] M. Haycock et.al., "3.2 GHz 6.4Gb/s per Wire Signaling in 0.18mm CMOS," *Intl. Solid-State Circuits Conf.*, pp. 62-63, 2001.

[12] R. Nair et.al., "A 28.5 GB/s CMOS Non-Blocking Router for Terabits/s Connectivity between Multiple Processors and Peripheral I/O Nodes," *Intl. Solid-State Circuits Conf.*, pp. 224-225, 2001.

[13] Y. Oowaki et.al., "A Sub-0.1µm Circuit Design with Substrate-over-Biasing," *Intl. Solid-State Circuits Conf.*, pp. 88-89, 1998.

[14] H. Banba et.al., "A CMOS Band-gap Reference Circuit with Sub-1V operation," *Symp. on VLSI Circuits*, pp. 228-229, 1998.

[15] S. Vangal et.al., "5GHz 32-bit Integer Execution Core in 130nm Dual-Vt CMOS," *Intl. Solid-State Circuits Conf.*, 2002.

Chapter 7

MEMORY LEAKAGE REDUCTION
SRAM and DRAM specific leakage reduction techniques

Takayuki Kawahara and Kiyoo Itoh
Central Research Laboratory, Hitachi Ltd.

7.1 INTRODUCTION

Low-voltage RAMs, especially embedded RAMs (e-RAMs) using nanoscale technology, are becoming increasingly important because they play critical roles in reducing power dissipation in the MPU/MCU/SoC for power-aware systems. Thus, research and development aimed at sub-1-V RAMs has actively been done, as exemplified by the recent development of a 0.6-V 16-Mb e-DRAM [1]. However, we face four major challenges to achieving such low-voltage e-RAMs [2-4]. These are: reducing the leakage current, maintaining the signal voltage and signal charge of RAM cells, reducing the speed variations caused by variations in MOSFET threshold voltage (V_T), and reducing the cell size. Of these, the leakage current issue is especially important because leakage loses the low-power advantages of CMOS circuits that we take for granted today. There are two major types of leakage. The first is sub-threshold current and the second is gate-tunneling current in MOSFETs, both of which increase rapidly when V_T and the gate-oxide thickness are reduced. Both types greatly affect the operation of RAM cells and peripheral circuits, not only in the standby mode but also in the active. Note that reducing sub-threshold current is up to the circuit designers, while reducing gate tunneling current is up to the process and device designers, as will be explained later.

To the best of our knowledge, almost all well-known basic circuit concepts to reduce sub-threshold current in low-voltage high-speed room-temperature operation LSIs had been proposed by 1993 mainly through developments in exploratory DRAMs [5-7]. The basic concepts of these proposals can be summarized as static and dynamic realizations of high-V_T or effectively high-V_T MOSFETs through various reverse-biasing schemes such as substrate (V_{BB}) reverse biasing [8], gate-source (V_{GS}) reverse biasing [9], or V_{GS} self-reverse biasing [10, 11]. The resulting high-V_T was proposed for use not only in RAMs but also in logic LSIs in the form of applications to all MOSFETs in a chip, a limited number of MOSFETs on non-critical paths in a chip [12], a power-switch (i.e., power gating) MOSFET [8], a buffer [9], various logic gates and an inverter chain [11], and iterative circuit blocks [10, 11]. All these applications were initially to reduce leakage in the standby mode. As early as 1993, however, they were quickly followed by attempts to reduce leakage in the active mode of a hypothetical 16-Gb DRAM [13, 14, 32]. Numerous attempts have subsequently been made to decrease leakage in logic LSIs [15], independently of RAMs, although RAM and logic designers must cooperate to develop reduction schemes for memory-rich LSIs. However, the problem of reducing sub-threshold current in the extremely high-speed active mode still remains unresolved.

The sub-threshold current issue is mainly described in this chapter in terms of circuit design. First, leakage and relevant issues with RAMs using nanoscale technology are discussed. Next, major leakage sources in RAMs, favorable features of RAMs that decrease this, and requirements to enable it are described. Then, basic concepts behind leakage reduction schemes that have been proposed to date for RAMs and logic LSIs are presented and compared, emphasizing the importance of gate-source reverse biasing schemes that enable rapid control and excellent leakage-reduction efficiency. After that, gate-source reverse biasing schemes will be discussed in detail, followed by applications to RAM cells and peripheral circuits in standby and active modes. Finally, future prospects are envisaged based on these considerations.

7.2 LEAKAGE IN RAMS

A major issue in producing low-voltage LSIs using nanoscale technology is how to reduce leakage current [2]. In particular, for RAMs, in addition to the leakage issue, we have to take two other relevant issues into consideration. These are signal voltage and signal charge issues with RAM cells, and variation issues with the V_T and V_T-mismatch between paired MOSFETs in flip-flop circuits such as SRAM cells and DRAM sense amps [2-4]. Otherwise, cell instability and unacceptably large variations in leakage as well as speed in RAM cells and peripheral circuits are likely to occur.

The intention of this section is to clarify these issues, which are common to both DRAMs and SRAMs. We have mainly assumed the RAM

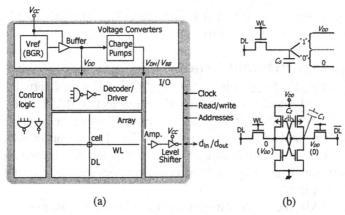

Fig. 7-1 Architecture of RAM chip (a) and RAM cells (b) with
DRAM cell for the upper and SRAM cell for the lower.

architecture in Figure 7-1 for this discussion. The chip is comprised of a
RAM cell array and peripheral circuits such as row/column decoders and
drivers, control logic circuits, amps and I/O circuits, and on-chip voltage
converters [2] that bridge the supply-voltage gap between the RAM cell
array and peripheral circuits.

7.2.1 Leakage Issues

The sub-threshold current in a MOSFET [2] is proportional to $W\,10^{(V_G - V_T)/S}$, where W is the channel width, V_G is gate voltage, V_T is threshold voltage,
and S is sub-threshold swing. This leakage is sensitive to V_G, V_T, and
temperature, which can easily be controlled by circuit techniques. For
example, for $S = 100$ mV/decade at 100°C, leakage is increased by one order
of magnitude with a V_T-decrement or V_G-increment of only 100 mV.
However it is not that sensitive to the device structure. Even a fully-depleted
(FD) SOI device reduces the S-factor by only about 20%. This implies that
developments with circuits are more effective in reducing leakage than those
with devices.

Gate tunneling current is very sensitive to gate oxide thickness (t_{OX})
while it is not that sensitive to V_G and temperature. For example, leakage is
reduced by one order of magnitude with a t_{OX} increment of only 2 to 3 A,
while the same reduction is attained with a V_G decrement of as much as 0.5
V in the low-V_{DD} region [16]. Since such large V_G control in the low-V_{DD}
region is risky in terms of stability, device developments are more effective
in reducing leakage than circuit developments. Thus, developments in new
gate-dielectric materials with low leakage and high-dielectric constant [17]
are urgently required as the ultimate solution. Even so, some circuits to
reduce leakage in peripheral circuits have been proposed, although they can
only be applied to the standby mode. They shut off the power supply path

through the insertion of a thicker-t_{OX} switch [18], and a gate floating circuit combined with a level keeper [16].

7.2.2 Leakage Relevant Issues

Signal Voltage and Signal Charge of RAM Cells: The read signal voltage of RAM cells and signal charge (Q_S) of non-selected RAM cells must be maintained to be sufficiently large to ensure stable cell-operation with reliable sensing and a small soft-error rate (SER) [2]. Unfortunately, they are degraded with device miniaturization, which always reduces cell-node capacitance and V_{DD}. They must be compromised with leakage (i) in cells, especially at a lower V_{DD}.

For DRAMs, the fresh voltage (i.e., V_{DD}) for a cell, just stored at the cell node with capacitance (C_S) by a full write, decays due to leakage and radiation-ray hitting. Thus, the signal voltage component in the cell is minimized at the maximum refresh time (t_{REFmax}) [2], which is guaranteed in catalog specifications. The resulting signal voltage (v_S) developed on the data line (DL) with capacitance (C_D) is expressed by effective signal charge ($= Q_S - Q_L - Q_C$) as $v_S = (Q_S - Q_L - Q_C)/(C_D + C_S)$, where Q_S is the original signal charge ($= C_S V_{DD}/2$), Q_L is the leakage charge ($= i\ t_{REFmax}$), and Q_C is the collected charge at the cell node due to alpha particle or cosmic-ray neutron hitting [2]. For successful sensing even at refresh operation, the signal voltage must be larger than the sum of various noises on the data line and the offset voltage of sense amps. However, Q_S is reduced with device miniaturization while offset voltage (i.e., V_T mismatch) is increased, as will be discussed later, although Q_C is usually reduced due to less collection area. Therefore, successful sensing is not ensured unless Q_L is greatly reduced. Unfortunately, however, t_{REFmax} must be longer with memory capacity to maintain the refresh busy rate. Thus, a greater decrease in leakage is necessary instead. Note that a smaller Q_S for given Q_L and Q_C also causes a larger SER with reduced effective signal charge.

For SRAMs, sub-threshold currents in a cell must be reduced to maintain the data retention current of SRAMs. However, this requirement prevents the V_T of cross-coupled and transfer MOSFETs from being scaled, and thus prevents V_{DD}-downscaling. For example, leakage of transfer MOSFETs must be reduced by using a higher V_T to eliminate interferences with other cells on the same data line. Leakage of cross-coupled paired MOSFETs must also be reduced with a higher V_T. Thus, the effective gate voltage for on-n-MOSFET, $V_{DD} - V_T - \delta V_T$, is rapidly reduced by the ever-lower V_{DD} and ever-larger δV_T (δV_T: V_T mismatch between paired MOSFETs) with device miniaturization. Consequently, this results in a smaller cell-read current and thus a smaller signal voltage on the data (bit) line. This is because the source voltage of the transfer MOSFET is raised more by a more degraded ratio for load and

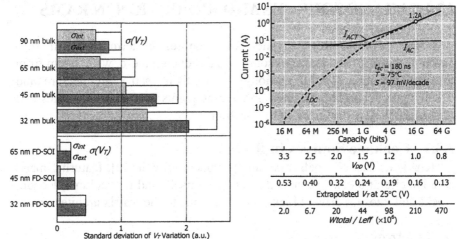

Fig. 7-2   Standard deviations of V_T variation, $\sigma(V_T)$, and intrinsic and extrinsic V_T variations, σ_{int} and σ_{ext}, respectively [45].

Fig. 7-3   Trends in active current (I_{ACT}), capacitive current (I_{AC}), and leakage current (I_{DC}) of DRAMs [2, 14, 21].

driver MOSFETs. The reduced effective gate voltage also narrows the static noise margin for non-selected cells, causing RAM cell instability. Moreover, reductions in node capacitances and V_{DD} increase SER because the soft-error critical charge needed to cause an error is reduced. Here, the critical charge is expressed by $(C_1 + 2C_2)\,V_{DD}$ using the cell-node capacitances (C_1) and node-to-node capacitance (C_2) [19] (Figure 7-1). Hence, on-chip ECC, an ever-larger additional C_1, or C_2 is needed [20].

_V_T and V_T-mismatch Variations:_ V_T-variations are expected to increase with device scaling, as we can see from Figure 7-2 [45]. This increase not only reduces the effective gate voltage of SRAM cells, but also causes variations in leakage (see Figure 7-11) as well as the speed of the chip [2-4].V_T-mismatch (δV_T) variations also increase since they are proportional to V_T variations. In particular, intra-die V_T and δV_T variations degrade the voltage margin of a chip because RAMs incorporate a huge number of tiny flip-flop circuits such as SRAM cells and DRAM sense amps. Note that the variations in SRAM cells are always larger, physically and statistically, than those in DRAM sense amps [2-4], because SRAMs must unavoidably use smaller cells with smaller MOSFETs, and there are a larger number of such flip-flop circuits in a chip. Thus, SRAMs will face the limitations with V_{DD}-downscaling earlier than DRAMs, because of the ever-decreasing effective gate voltage with higher V_T and δV_T, and lower V_{DD}.

7.3 LEAKAGE SOURCES AND REDUCTION IN RAMS

RAMs offer inherent advantageous features in terms of their sub-threshold-current (i.e., leakage) sources and reducing these, which stem from physical and circuit configurations due to their matrix architecture and circuit operations. The features and requirements to reduce sub-threshold current are briefly explained in this section.

7.3.1 Leakage Sources in RAMs

RAMs consist of a cell array and peripheral circuits [2], featuring many iterative circuit blocks, such as the cell array itself, and row/column decoder blocks and relevant driver blocks, which are due to the matrix architecture.

(1) Cell Array
The memory cell array is the largest for leakage sources of the iterative circuit blocks because it has the largest number of cells despite the small channel width of each cell. Leakage in RAM cells causes different detrimental effects for DRAM and SRAM cells. Increased leakage in the cell storage node of a DRAM cell shortens the data retention time (i.e., the refresh time). However, leakage dramatically increases the data retention current of an SRAM cell along with decreasing V_T. The use of a high-V_T or effectively high-V_T MOSFET is eventually the solution, as will be explained later.

(2) Peripheral Circuits
Iterative circuit blocks are major sources of leakage in peripheral circuits, while the remaining circuits are minor. This is because each block usually has a large total channel width involving leakage, and thus the leakage accumulated from many iterative circuit blocks becomes larger with decreasing V_{DD} and thus V_T. For example, as we can see from Figure 7-3 [2, 14, 21], the DC sub-threshold leakage (I_{DC}) from such iterative circuit blocks in DRAM chips rapidly increases, and surpasses AC capacitive current (I_{AC}), eventually dominating the total chip current (I_{ACT}). Note that in some cases leakage reduction of p-MOSFETs is more important than that of n-MOSFETs because of the larger channel width and sometimes larger S-factor of p-MOSFETs ([22] or see Figure 7-17).

Array Driver Blocks: Of the iterative circuit blocks, the row (i.e., word) driver block contributes most to leakage because it has the largest total channel width. Note that each p-MOSFET word driver has a large channel width to quickly drive a heavily capacitive word line that is composed of many gate capacitances of cells. Here, a p-MOSFET is necessary for the output driver to simplify the design with reduced stress voltage to the MOSFET [2]. The total channel width of the column driver block is usually narrow because each driver theoretically controls only one column switch when connecting a selected DL to a common DL. Even with a multi-divided data-line architecture [2] the number of switches is limited.

Row and Column Decoder Blocks: The total channel width is effectively narrow because each decoder block consists of robust circuits (i.e., leakage-immune NAND gates) [2].

Sense Amp Block: The block is another concern, especially in DRAMs. The design of sense amps (SAs) [2, 3], which usually have a cross-coupled circuit configuration to provide low power and a small area, can be different for DRAMs and SRAMs. This is because the necessary size, the number in a chip, and circuit operation are usually different. DRAMs feature a huge number of tiny SAs in a chip, because one SA must be placed at each data line due to refresh requirements. In addition, in standard mid-point (half-V_{DD}) sensing of DRAMs, the SA must operate at the lowest voltage (i.e., half-V_{DD}) in the chip, calling for a lower V_T for fast sensing. The many low-V_T SAs resulting from this thus generate large leakage despite small MOSFETs being used. In contrast, SRAMs have a small number of SAs, although they must be highly sensitive to achieve high speed.

Control Logic Block: This block may operate at its highest frequency, making leakage reduction extremely difficult. Fortunately, however, the total channel width of the block is narrow (usually 10% at most of total channel width of chip). The I/O block operating at high I/O power supply voltage accepts a sufficiently high V_T to suppress leakage.

7.3.2 Features of RAMs

The leakage current can drastically be reduced by utilizing some of following distinctive features of RAMs [2, 3].

Use of Multiple Iterative Circuit Blocks: RAMs consist of multiple iterative circuit blocks, as previously discussed. In addition, even in the active mode, all circuits in each block, except for the selected one, are inactive. This feature is extremely important to reduce the leakage in each block, simply and effectively, even in the active mode with a smaller area penalty than logic LSIs, as will be explained later.

Use of Input-predictable Circuits: RAMs are almost composed of input-predictable circuits, allowing circuit designers to predict all node voltages in the chip and to prepare the most effective leakage reduction scheme in advance.

Use of Robust NAND Gates: Modern CMOS RAMs do not use leakage-sensitive circuits, such as dynamic NOR gates, that require a level keeper to prevent malfunctions caused by leakage [24]. For example, the decoder block consists of dynamic (for the row) and static (for the column) NAND gates to reduce power [2]. The NAND-decoder block discharges only one output node in a selected decoder, while the NOR-decoder block used in the n-MOS era discharges all output nodes in the block, except for the selected one. In addition to low power, they reduce leakage through the stacking effect [3], as will be explained later. In the standby mode, RAMs have a feature in that all address inputs to each n-MOSFET NAND decoder are low, enabling leakage to be dramatically reduced. Even in the active mode, leakage is reduced with the effect because a considerable number of decoders have at least two low-level inputs.

Slow RAM Cycle: RAMs feature a slow cycle compared with random logic-gate LSIs, and this allows each circuit to be active for only a short period within the "long" memory cycle, leaving additional time to control leakage. This is true for DRAM row circuits, which are sufficiently slow to accept leakage control. However, the column circuits in modern DRAMs [2, 3] feature a fast burst cycle and unpredictable circuit operation (every column may be selected during the memory cycle). Therefore, it is difficult to reduce leakage in column circuits in the active mode. Such is the case for extremely high-speed SRAMs and logic LSIs. Fortunately, however, the total channel width of such circuits in RAMs is narrow, as previously discussed.

Utilization of Internal Power-supply Voltages: Multi-static V_T [2, 3] is available for DRAMs because the internal power-supply voltages necessary for DRAM operations can be utilized to achieve a high V_T.
Use of Huge Number of Tiny Flip-flop Circuits: This is a disadvantage with RAMs. This is because flip-flop circuits for SRAM cells and DRAM sense amps inherently have narrow voltage margins because of the necessity for high and unscalable V_T MOSFETs, and the operations are susceptible to V_T-mismatch variations.

7.3.3 Requirements for Leakage Reductions

A high-speed reduction scheme for the active mode will eventually have to be developed [4] although this is more difficult than for the standby mode since leakage needs to be controlled much faster. Once such a high-speed scheme is available it can also be applied to the standby mode, making its design much easier as it can be applied to both. Even if the scheme can be applied to the standby mode, it may not be able to be applied to the active mode if it is too slow, necessitating a complicated design with two different reduction schemes. Note that such a high-speed scheme can also do mode transitions at high speed while reducing leakage, as exemplified by a fast recovery time from standby to active mode, which is the key to mobile applications. The key to achieving reduction is to control leakage within one active cycle so that the leakage in each block is reduced in an active cycle, ready to reduce the leakage of any block in the next active cycle. Hence, better reduction efficiency, allowing a smaller voltage swing for a given leakage reduction, smaller load capacitance, and simpler control are required.

In addition to high-speed, confining peripheral circuits to a minimum active circuitry is vital. Leakage in the active mode can further be reduced, if the scheme enables active (i.e., selected) circuitry in each block to be confined to the small every instance within a cycle time [4, 13, 32], and reduces the leakage in the remaining inactive (i.e., non-selected) circuitry that dominates the total leakage in the block. Fortunately, peripheral circuits in RAMs accept such schemes and reduce leakage easily and effectively, whereas random logic circuits in MPU/MCU/SoC do not, as will be discussed later. This stems from the features of RAMs we described. Moreover, the area penalty must be minimized. Furthermore, compensation for leakage variations as well as speed variations caused by V_T-variations is essential. This will also improve the voltage margin of the worst SRAM cell with the highest V_T and the largest V_T-mismatch.

7.4 VARIOUS LEAKAGE REDUCTION SCHEMES

Leakage reduction schemes for MOSFETs are described in general in this section, and then compared in terms of RAM designs [3].

7.4.1 Leakage Reductions in MOSFETs

Increasing V_T is the best way to reduce the sub-threshold current (i) of a MOSFET. This is expressed by

$$i \propto \exp\left[\pm\frac{V_{GS} - V_T - K(\sqrt{V_{BS} + 2\Psi} - \sqrt{2\Psi}) + \lambda V_{DS}}{S/\ln 10}\right] \times \left\{1 - \exp[-\frac{qV_{DS}}{kT}]\right\} , \quad (1)$$

where plus values refer to n-MOSFETs and minus values to p-MOSFETs. V_T is the actual threshold voltage, while S is the sub-threshold swing, K is the body-effect coefficient, and λ is the drain-induced barrier lowering (DIBL) factor [25]. Here, q is the electronic charge, k is the Boltzmann constant, and

Table 7-1   Concepts to create effective high V_T [3].

		Modified voltage(s)	NMOST	PMOST
(A) V_{GS} reverse biasing	(A1) V_S: self-reverse biasing			
	(A2) V_G: offset gate driving			
(B) V_{BS} reverse biasing	(B1) V_B: substrate driving			
	(B2) $V_S = V_G$: offset source driving			
(C) V_{DS} reduction				

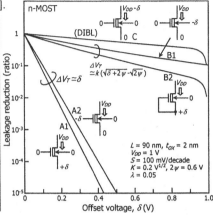

Fig. 7-4   Leakage reduction efficiency [3].

T is the absolute temperature. Leakage is usually reduced to 1/10 with a V_T increment of only 0.1 V for a bulk MOSFET with $S \cong 100$ mV/decade at 100°C.

Two ways of obtaining a high-V_T MOSFET from a low-actual-V_T MOSFET are by increasing the doping level of the MOSFET substrate and by applying reverse biases. Thus, the selective use of these resulting high-V_T MOSFETs in low-actual-V_T circuits or the reverse biasing of low-actual-V_T circuits decreases circuit sub-threshold current. Although there have been many attempts to develop reverse-biasing schemes, the basic concepts can still be categorized into the three in Table 7-1. These are (A) gate-source (V_{GS}) reverse biasing, (B) substrate-source (V_{BS}) reverse biasing, and (C) drain-source voltage (V_{DS}) reduction. Here, the V_{GS} reverse biasing scheme can further be categorized as V_S-control with a fixed V_G (A1) [10, 11] and V_G-control with a fixed V_S (A2) [9]. The V_{BS} reverse biasing schemes can be categorized as V_B-control with a fixed V_S (B2) [8, 33] and V_S-control with a fixed V_B (B2) [34, 35].

The efficiencies for reducing leakage for offset voltage δ applied to a low-actual V_T MOSFET are plotted in Figure 7-4 using 0.1-μm MOSFET parameters. The reduction efficiency of (A2) is the leakage (i)-ratio without and with V_{GS} reverse bias:

$$r_1 = \frac{i(V_{GS} = 0)}{i(V_{GS} = -\delta)} = \exp(\frac{\delta}{S / \ln 10}) \qquad (2)$$

This is quite large because δ has been directly added to the low-actual V_T. The reduction efficiency of (B1) is calculated in the same manner:

$$r_2 = \exp[\frac{K(\sqrt{\delta + 2\Psi} - \sqrt{2\Psi})}{S/\ln 10}] \qquad (3)$$

This is smaller than r_1 because of the dependence of square root on δ and the small K. (C) has quite a small reduction efficiency of

$$r_3 = \exp(\frac{\lambda\delta}{S/\ln 10}) \qquad (4)$$

because of the small λ, unless V_{DS} approaches thermal voltage (kT/q), where i is drastically reduced as the second factor of Eq. (1). Scheme (A1) has the largest reduction efficiency of $r_1r_2r_3$ because all three effects are combined. (B2) has a reduction efficiency of r_2r_3, which is larger than that of (B1) because of the additional effect of reducing V_{DS}. Note the inherently small offset voltage required to reduce the given leakage provided by scheme (A). This effectively reduces not only sub-threshold current in the low-power mode, but also achieves a faster recovery time in the high-speed mode, as was explained earlier.

The concept involves two types of biasing, static and dynamic. The former, or so-called dual-V_T scheme, is used to statistically combine low-V_T MOSFETs and the resulting high-V_T MOSFETs in core circuits. A CMOS dual-V_T scheme [26, 27] in which a low V_T is applied only to the critical path occupying a small portion of the core is quite effective in simultaneously achieving high speed and low leakage current, although the basic scheme was proposed for an n-MOSFET 5-V 64-Kb DRAM [28]. A difference in V_T of 0.1 V reduces the standby sub-threshold current to one-fifth its value for a single low V_T, although an excessive V_T difference might cause a race condition problem between low- and high-V_T circuits. The dual-V_T scheme is also applied to SRAMs [27, 29]. It has been reported that a combination of dual V_T and dual V_{DD} achieved a high-speed low-power 1-V e-SRAM [29]. Another application of the dual-V_T scheme is in a high-V_T power switch [8, 10-12, 14, 15] that can cut the sub-threshold current of an internal low-V_T core in the standby mode. High-V_T MOSFETs can easily be produced in DRAMs [22] by using the internal supply voltages that DRAMs require, as will be explained later. The high V_T, however, eventually restricts the lower limit of V_{DD} as the transconductance of a MOSFET degrades at a lower V_{DD}. The latter (i.e., dynamic biasing) changes the V_T so that it is sufficiently low in high-speed modes, such as the active mode with no reverse bias, while in low-power modes, such as the standby mode, it is increased by changing bias conditions (see Table 7-1).

7.4.2 Comparisons of Concepts

There is a vast difference between the two schemes (A and B) in terms of recovery time [3]. In V_{GS} reverse biasing, the small voltage swing, δ, enables quick recovery (several nanoseconds for RAMs). In V_{BS} reverse biasing, however, it takes more than 100 ns for recovery when it is applied to a power line with a heavy capacitance, because V_{BS} reverse biasing requires a large V_B swing (ΔV_B) or V_S swing (ΔV_S), which is usually more than 1.5 V for a given change in V_T (ΔV_T). The necessary voltage swing imposes different requirements on substrate driving (B1) and offset source driving (B2). In (B1), the necessary voltage is significantly larger than V_{DD}, which is the sum of V_{DD} and ΔV_B. For example, existing MOSFETs with a 0.2-V$^{1/2}$-body-effect coefficient (K) require a ΔV_B as large as 2.5 V to reduce the current by two decades with a 0.2-V ΔV_T. A larger-K MOSFET is needed to reduce the swing. However, this slows down the speed in stacked circuits, such as NAND gates. In contrast, the K value decreases with MOSFET scaling, implying that the necessary ΔV_B will continue to increase further in the future (i.e., unscalable ΔV_B) due to a lower K, and there will be a need for a larger ΔV_T reflecting the low-V_T era. Eventually, this will enhance short-channel effects and increase other leakage current, such as GIDL current [30]. A shallow reverse V_B setting, or even a forward V_B setting in active mode, is also required to effectively increase V_T in standby mode, because V_T is more sensitive to V_B [2]. However, the requirements to suppress V_B noise will instead become more stringent. In fact, a connection between the substrate and source every 200 μm [31] to reduce noise has been proposed, despite the area penalty. In addition, problems inherent in LSIs with an on-chip substrate bias (V_{BB}) generator, which DRAM designers have experienced since the late 1970s, may occur even with a low V_{DD}. These problems include spike current and CMOS latch-up during power-on and mode transitions, and V_{BB} degradation caused by increased substrate current in high-speed modes and screening tests at high stress V_{DD}. They also include slow recovery time as a result of poor current drivability of the on-chip charge pump.

In offset source driving (B2), the necessary voltages and voltage swing at any node are smaller than V_{DD}. This control becomes ineffective as V_{DD} is lowered owing to a smaller substrate bias. However, these problems accompanied by an on-chip V_{BB} generator are not expected. The energy overhead for offset source driving (B2) through mode transitions is usually larger than that for substrate driving (B1). This is because the parasitic capacitances of source lines are larger than those of substrate lines, though the necessary δ is smaller. The parasitic capacitances of substrate lines consist mainly of junction capacitances between the substrate (well) and source/drain of MOSFETs, while those of source lines include the gate

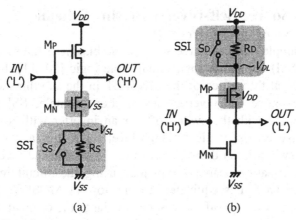

Fig. 7-5 SSI applications to inverter with low input (a) and high input (b) [10, 11].

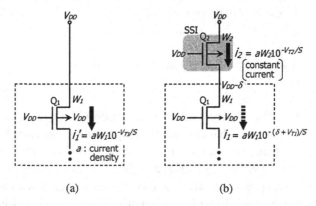

Fig. 7-6 Circuit without SSI (a) and with SSI (b) [10, 11].

capacitances of on-state MOSFETs as well as junction capacitances. The energy overhead for self-reverse biasing (A1) is quite small because of the small and self-adjusted δ.

7.5 GATE-SOURCE REVERSE BIASING SCHEMES

Gate-source reverse (or back) biasing (i.e., A in Table 7-1) is discussed in detail in this section because it is the most promising for RAM designs. It can be categorized as gate-source self-reverse biasing (A1), and gate-source offset driving (A2). In particular, self-reverse biasing is most effective in RAM designs because of its inherent features. These are fast and simple control that can be applied even to the active mode, a small area penalty, and confinement to minimum active circuitry if applied to an iterative circuit block in RAMs.

7.5.1 Gate-Source Self-reverse Biasing Scheme

(1) Switched-Source Impedance (SSI)

A typical example of gate-source self-reverse biasing is the switched-source impedance (SSI) applied to input predictable logic [10, 11]. In Figure 7-5, SSI is stacked at the source of the MOSFET in the sub-threshold region to provide gate-source self-reverse biasing. For example, SSI is at the n-MOSFET source with the switch off for an inverter with low-level input during inactive eriods, while for high-level input it is at the p-MOSFET source. The switch is on during active periods so that the inverter operates normally. SSI features can be explained using the circuit in Figure 7-6, assuming that an SSI is equivalent to one low-V_T MOSFET (Q_2). During inactive periods with Q_2 off, no matter how large the original sub-threshold current (i_1') is, it is eventually confined to the Q_2 constant current (i_2) with self-adjusting δ. The δ and leakage reduction ratio γ ($= i_1/i_1' = i_2/i_1'$) are simply expressed by making i_1 equal i_2 as

$$i_1 = a\ W_1\ 10^{-(\delta+V_{T1})/S}, a: \text{current density,} \tag{5}$$
$$i_2 = a\ W_2\ 10^{-V_{T2}/S}, \tag{6}$$
$$\delta = (V_{T2} - V_{T1}) + (S/\ln 10)\ \ln(W_1/W_2), \text{ and} \tag{7}$$
$$\gamma = 10^{-\delta/S}. \tag{8}$$

Primary Stacking Effect: Additional MOSFET Q_2 makes δ and thus γ adjustable with its channel width (W_2) and/or threshold voltage (V_{T2}), yielding the primary stacking effect. A larger δ and thus a smaller γ are attained with a smaller W_2 and/or a higher V_{T2}. In an extreme case where V_{T2} is high so that i_2 becomes almost zero, γ is almost zero with a large δ. This is just the power switch with sufficiently high V_T. Therefore, there is no substantial difference between the power switch and 'leaky' power switch with a lower V_T, although a higher V_T reduces leakage more, but needs a longer recovery time instead. For the same V_T, γ is simply expressed as the channel-width ratio of W_2/W_1. For example, a small voltage swing (δ) of only 0.1 V reduces leakage by one order of magnitude when $W_2 = W_1/10$ and $S = 100$ mV/decade. Hence, fast control is achieved with such small δ, small load capacitance of the node, and simple self-reduction control.

Secondary Stacking Effects: If $W_2 = W_1$ and $V_{T2} = V_T$, no reduction in leakage is expected because of $\delta = 0$ and $\gamma = 1$. It is true as long as the above equations, where secondary effects have been neglected, are used. However, the secondary effects deriving from offset source driving (B2 in Table 7-1) to Q_1, and the DIBL effect (C in Table 7-1) of Q_2 actually reduce leakage although these reductions are quite small, as we can see from Figure 7-4. As a first approximation, leakage is at least halved because Q_1 and Q_2 are equivalent to one MOSFET with halved channel width. To be more exact, about one-third reduction is expected [3].

Many applications of SSI to logic gates [11], such as inverters, NAND gates, NOR gates, clocked inverters, and R-S latches have been proposed.

(2) SSI Sharing

One of the most effective SSI applications is SSI sharing [2-4, 10, 11, 13, 14, 32]. Figure 7-7 shows examples of SSI sharing applied to an inverter chain and an iterative circuit block to minimize the area penalty. For a 0-V input inverter chain, in which all outputs of each inverter have settled, leakage flows from the V_{DD} supply to a p-MOSFET in each V_{DD} input inverter. Thus, there is accumulated leakage flowing into the p-MOSFET SSI so that it is reduced with resulting δ. The accumulated leakage is proportional to the total channel width of relevant p-MOSFETs in the inverter chain. Thus, the leakage reduction ratio is expressed by the ratio of the channel width (W) of p-MOSFET SSI to the total channel width of the p-MOSFETs, i.e., $\gamma = W/\Sigma W_k$ ($k = 1$ to n) for a 2n-stage inverter chain. Here, the W can be comparable to each p-MOSFET channel width without a speed penalty, because each inverter switches with different timing during active periods. Thus, γ is eventually expressed as $1/n$. Therefore, with a large n, leakage and area penalties become smaller, because γ is almost 0, and W is much narrower than the total channel width. As a result, the leakage in the inverter

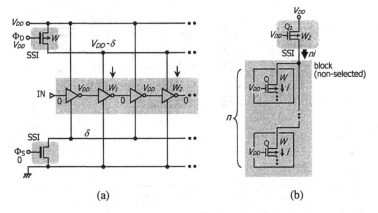

(a)   (b)

Fig. 7-7 SSI sharing for inverter chain (a) and iterative circuit (b) [11].

chain can be reduced to that of p-MOSFET SSI. Such is the case for n-MOSFET SSI. The n iterative inverter block (b) with each represented by one p-MOSFET, coupled with SSI connected to the common source line, also minimizes the area penalty. During non-selected periods at the V_{DD} gate voltage of each MOSFET, the block is equivalent to one p-MOSFET with W_1 equal to nW, so that the leakage reduction ratio (γ) is simply expressed as W_2/nW. Here, W_2 can be comparable to Q-channel width (W) without a speed penalty, because only one MOSFET is activated during selected periods with SSI on. Thus, γ is eventually expressed as $1/n$. Therefore, the leakage and area penalties are greatly reduced with a large n.

(3) SSI Applications to Multi-divided Block

Confinement to a minimum active circuitry further reduces leakage in the active mode, if the large leakage from the remaining large inactive circuitry is sufficiently reduced by SSI. This can be done by partially activating the multi-divided block using SSI. Table 7-2 compares various SSI applications to an iterative circuit block [13, 14, 32]. Configuration (b) corresponds to Figure 7-7(b), (c) is a multi-divided block with m sub-blocks with n/m circuits each, coupled with SSI connected to the common source to select each sub-block, and (d) is hierarchical SSI with additional SSI (i.e., SSI$_1$ and SSI$_2$). Here, the same V_T, the same S, and the same leakage for p-MOSFETs and n-MOSFETs have been assumed. The channel-width ratios of the SSI MOSFET and the additional SSI MOSFET to the circuit MOSFET that flows leakage (i) with a channel width of W have also been assumed to be a and b, respectively. The leakage currents for configurations (b), (c), and (d) in the standby mode are confined to their SSI constant currents, ai, mai, and

Table 7-2   Various SSI applications to iterative circuit block [13, 14, 32]

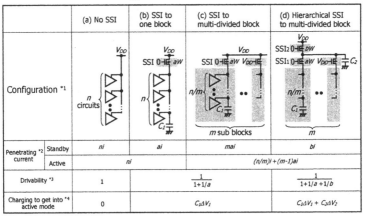

Configuration [1]	(a) No SSI	(b) SSI to one block	(c) SSI to multi-divided block	(d) Hierarchical SSI to multi-divided block
Penetrating [2] current — Standby	ni	ai	mai	bi
Penetrating [2] current — Active	ni		$(n/m)i + (m-1)ai$	
Drivability [3]	1		$\dfrac{1}{1+1/a}$	$\dfrac{1}{1+1/a+1/b}$
Charging to get into [4] active mode	0		$C_1\Delta V_1$	$C_1\Delta V_1 + C_2\Delta V_2$

[1] Gate voltages shown are for the active mode. p-MOSTs in iterative circuits have a gate width W.
[2] All the transistors have the same threshold voltage. Channel widths are tailored as $a \ll n$ and $b \ll ma$.
[3] Relative value to the conventional ignoring wiring resistance.
[4] C_1, C_2, and C_3 wiring capacitances and ΔV_1, ΔV_2, and, ΔV_3 are voltage drops from V_{DD} in the standby mode.

bi, respectively, no matter how large the total circuit leakage and the total SSI leakage are. In the active mode, leakage (*ni*) flows for (b) with SSI on. Leakage (*n/mi* + (*m*-1) *ai*) flows for multi-divided block (c) because one selected SSI and one selected circuit in the selected subblock are turned on, while other SSIs and other circuits remain off. For another multi-divided block (d), i.e., hierarchical SSI, the same leakage flows with SSI_2 on. The standby leakage is minimized by architecture (d) because the *mai* current of (c) is eventually confined to *bi*, and active leakage is reduced by multi-divided blocks (c) and (d). Active leakage is minimized by the condition of $m = \sqrt{n/a}$ for $m \gg 1$, although this reduction must be compromised with speed.

(4) Compensations for V_T-variations
Whatever SSI scheme is applied, the constant leakage current expressed by Eq. (6) still remains. It strongly depends on V_T and temperature, and thus compensation for variations is the key, as will be discussed later.

7.5.2 Gate-Source Offset Driving Scheme

A negative supply voltage (δ) or a supply voltage (V_{DH}) higher than V_{DD} is essential to substantially cut a low-V_T n-MOSFET or p-MOSFET. This is because a high-V_T n-MOSFET is effectively achieved with the sum of δ and a low-actual V_T, or a high-V_T p-MOSFET is effectively achieved with the sum of ($V_{DH} - V_{DD}$) and a low-actual V_T. A scheme that corresponds to A2 in Figure 7-4 has been widely used in modern DRAMs [2], and will be discussed later. Such a scheme has also been proposed for power switches because it provides a large ON-current. However, there are many challenges to ensuring stable operation, especially in high-speed power switches.

Gate-voltage Setting Accuracy: The gate voltage of a power switch MOSFET must be well regulated against random high-speed operations of internal circuits, because the leakage in a MOSFET is quite sensitive to gate voltage. For example, the accuracy of setting the gate voltage must be less than 30 mV for allowable leakage variations of less than 50 % assuming $S = 100$ mV/decade. Unfortunately, gate voltage (e.g., negative or raised supply voltage) is usually generated by the on-chip charge pump with inherently poor current drivability [2], which causes unregulated gate voltage. Moreover, the quasi-static output level is susceptible to various coupling noises. These problems are similar to those in the on-chip V_{BB} generator that was described earlier.

Active Mode: The load current of the converter (e.g., on-chip charge pump) is maximized at the highest frequency and highest voltage of the load, while it is minimized at the lowest frequency and lowest voltage. The output voltage (i.e., gate voltage of the power-switch MOSFET) of the converter

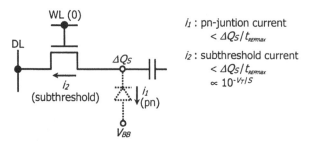

Fig. 7-8 Two leakage-current components of non-selected DRAM cell [2].

must be sufficiently precise against such wide load-current variations. Here, a large load current increases the power dissipation of the converter since the current must be compensated for, e.g., by a high pumping frequency to stabilize the output voltage. Thus, the smallest load capacitance possible is preferable to stabilize the output voltage with reduced load current.

Standby Mode: The DC currents of the load such as leakage must be sufficiently small to keep the output voltage constant after the converters have been switched to low-power modes to minimize the driving current. This is essential when an on-chip charge pump is used.

Other Modes: Possible problems [2, 3] include spikes or large leakage current during power-on or during mode transitions. Such abnormal currents may occur while on-chip voltages have not yet settled. In addition, the output voltage of converters may change due to the slow recovery time involved with the on-chip charge pump. In addition, voltage degradation may be caused by increased load current in screening tests at high stress V_{DD}, as previously discussed.

7.6 APPLICATIONS TO RAM CELLS

Leakage-conscious designs for RAM cells [3, 4, 36] are described in this section, which include discussions on the necessary V_T in RAM cells to reduce sub-threshold leakage, the necessary word voltage and power supply, and typical leakage reduction schemes that have been proposed to date.

7.6.1 DRAM Cells
The leakage in a DRAM cell flowing from the cell storage node to the data line shortens the refresh time. There are two leakage currents at the storage node of a non-selected cell [2]: the p-n junction leakage current (i_1) to the substrate, and the sub-threshold current (i_2) to the data line (DL), as we can see in Figure 7-8. Even with such leakage currents, each non-selected

cell must hold data for the t_{REFmax} (i.e., the maximum necessary refresh time). The p-n current can be estimated by measuring the date retention time (so-called static refresh time), while applying a static V_{DD} to the data line and 0 V to the word line so that the sub-threshold current to the data line is eliminated. The sub-threshold current is estimated by measuring the data retention time under conditions where 0 V is applied to the data line as long as possible. In practice, the conditions are achieved by a set of successive low-level ("L") data-line disturbances that is done with successive operations of other cells on the data line [2]. The refresh time measured under the disturbances is called the dynamic refresh time. The sub-threshold current component becomes negligible if cell V_T is sufficiently high. A longer t_{REFmax} needs a higher V_T.

The lowest necessary V_T ($=V_{TLD}$) for a MOSFET depends on the necessary t_{REFmax}. The V_{TLD} is defined as the V_T by which the sub-threshold current, from the cell node to the data line during "L" data-line disturbances as previously described, is suppressed to the extent of satisfying t_{REFmax} specifications. The t_{REFmax} depends on the row cycle time (t_{RC}), the refresh busy rate η ($= n\ t_{RC}\ /\ t_{REFmax}$), and the maximum junction temperature [2]. Here, n and t_{RC} denote the refresh cycles (i.e., the number of rows in the logical array configuration) and the row cycle time. A shorter t_{REFmax} that allows a larger sub-threshold current accepts a lower V_T, enabling low-V_{DD} operation. In general, the V_{TLD} must be quite high and unscalable [4] to satisfy the t_{REFmax} requirement. This high V_{TLD} calls for a high word-line voltage (V_W) and a thick-t_{OX} MOSFET, since the V_W

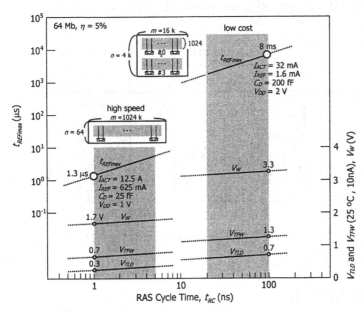

Fig. 7-9   Maximum necessary refresh time (t_{REFmax}) and necessary cell V_T [4].

must be higher than the sum of V_{DD} and V_{TFW} to do a full-V_{DD} write [2]. Here, V_{TFW} is the sum of V_{TLD} and the V_T-increase that is due to the body effect developed by raising the source (i.e., storage node) by V_{DD}.

To achieve low-cost, the V_{TLD} is higher than that for a high-speed design because a longer t_{REFmax} is required. The requirement for a small chip, which is achieved by a large number of cells connected to a data line, results in large data-line capacitance (C_D), and thus a long t_{RC} and a high V_{DD} for sufficient signal voltage. For a given refresh busy rate, the long t_{RC} necessitates a longer t_{REFmax} (i.e., less leakage to the data line) and thus a high V_{TLD}. For example, if a low-cost hypothetical 64-Mb DRAM [4] with a pair of data lines connecting 1,024 cells can operate at $C_D = 200$ fF, $t_{RC} = 100$ ns, and $V_{DD} = 2$ V, the calculated values for t_{REFmax}, V_{TLD}, V_{TFW}, and V_W are 8 ms, 0.7 V, 1.3 V, and 3.3 V, respectively, as we can see from Figure 7-9. However, if a hypothetical high-speed 64-Mb DRAM with a pair of data lines connecting 64 cells can operate at $C_D = 25$ fF, $t_{RC} = 1$ ns, and $V_{DD} = 1$ V, these values are 1.3 μs, 0.3 V, 0.7 V, and 1.7 V. Here, we have assumed $\eta = 5$ %, $C_S = 30$ fF, body effect $K = 0.5$ V$^{1/2}$, acceptable decay ($\Delta Q_S / Q_S) = 0.1$, leakage $i = \Delta Q_S / t_{REFmax}$, $S = 120$ mV/decade (100°C), and $V_T(25°C) = V_T(100°C) + 0.15$ V.

Even with a low-actual V_T, a high V_T (i.e., V_{TFW}) is effectively achieved as the sum of the negative voltage applied to non-selected word lines and a low-actual V_T. This is the so-called negative word-line (NWL) scheme and produces gate-source offset driving (i.e., A2 in Tab. 7-1 or Figure 7-4). The NWL scheme is better than word bootstrapping because the MOSFET gate stress voltage during activation is relaxed [2].

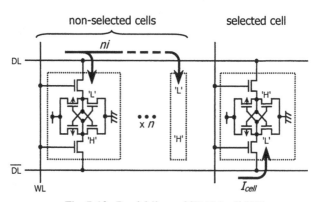

Fig. 7-10 Read failure of SRAM cell [23].

7.6.2 SRAM Cells

A low-V_T transfer MOSFET may cause read failure [23], if the sub-threshold current (ni) accumulated from the transfer MOSFETs of many non-selected cells along the same data line is larger than the read current (I_{cell}) of the selected cell (see Figure 7-10). This condition restricts the number of cells connected to one data-line pair, affecting the array configuration of the SRAM. A sufficiently high V_T, however, reduces the sub-threshold current of the transfer MOSFETs. The leakage from cross-coupled MOSFETs dramatically increases the data retention current in the cell along with decreasing V_T, as exemplified by the 1-Mb array current in Figure 7-11 [4]. Therefore, there is a minimum V_T for the 6-T SRAM cell to satisfy retention-current specifications [2-4]. If a low-power 1-Mb SRAM accepts a leakage of 0.1 μA at T_{jmax} = 75°C, the V_T at 25°C might be as high as 0.71 V, while if a high-speed 1-Mb SRAM allows a leakage of 10 μA at T_{jmax} = 50°C, the V_T can be as low as 0.49 V. Such high-V_T MOSFETs, however, degrade the signal charge and static noise margin. This eventually prevents V_{DD}-downscaling, especially for a low-power SRAM necessitating a high-V_T. Note that the leakage in an array that is actually determined by the mean value (V_T(mean)) of intra-die V_T variations varies with V_T(mean) and temperature. For example, it varies by as much as about four orders of magnitude (from point C to F in the figure) if we assume the V_T variation = ± 100 mV and the junction temperature variation = 100°C (25 to 125°C) for V_T (mean) = 0.49 V. Thus, chip-to-chip compensations for leakage variations are essential.

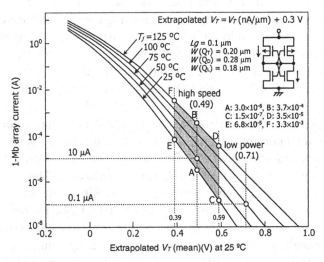

Fig. 7-11   Data-retention current of 1-Mb array versus V_T of cross-couple MOSTs [4].

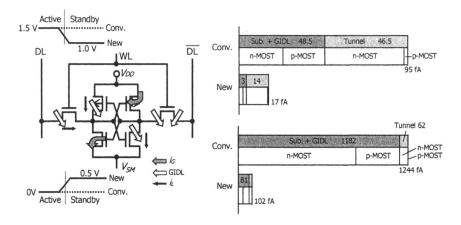

Fig. 7-12 Leakage current reduction for low-power SRAM [37].
i_G and i_L: gate tunneling current and subthreshold current.

Dynamic High V_T: A new driving scheme (Figure 7-12) to reduce leakage has been proposed and applied to a 1.5-V, 27-ns access, 6.42 x 8.76 mm^2, 16-Mb SRAM [37]. This is an application of B2 in Table 7-1 to the SRAM to achieve a dynamic high-V_T cell. Note that the gate and source are connected through the on-MOSFET. The scheme lowers the data-line voltage from 1.5 V to 1 V and raises the ground line to 0.5 V during the transition from active to standby mode, enabling the total leakage current per cell in the standby mode to be reduced. At ambient temperature, the measured total current of the conventional scheme was 95 fA. The largest component was the sum of sub-threshold current and GIDL current for the n-MOSFET and p-MOSFET, although the V_Ts were as large as 0.7 V and 1 V, respectively. The gate-tunneling current of the n-MOSFET was comparable to this, despite an electrical t_{OX} as thick as 3.7 nm. The scheme greatly reduced the total current to 17 fA. Offset source driving by 0.5 V applied to the driver and transfer n-MOSFETs, and relaxing the electric field of all MOSFETs by 0.5 V were responsible for the reduction. The reduction was more remarkable at a higher temperature. At 90°C, the total current of conventional scheme was drastically increased to 1,244 fA because of the increase in the sub-threshold-current component. Note that GIDL current and gate-tunneling current are insensitive to temperature. The scheme reduced the total current to 102 fA. To cope with the increased SER caused by the reduced signal charge in the standby mode, an ECC was incorporated with a speed penalty of 3.2 ns and an area penalty of 9.7%.

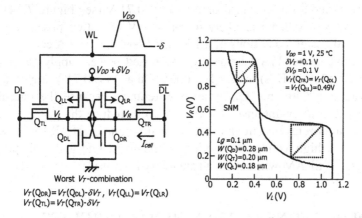

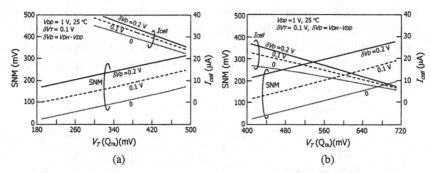

Fig. 7-13   Raised power supply 6-T SRAM cell, (a) and butterfly curve (b) [4, 36, 38].

Static High-V_T: Figure 7-13 shows the raised power-supply ($V_{DH} = V_{DD} + \delta V_D$) and dual-$V_T$ cell scheme [4, 36, 38]. Static high-V_T cross-coupled MOSFETs are used to reduce the sub-threshold currents of cross-coupled MOSFETs. Low-V_T transfer-MOSFETs, coupled with the NWL scheme (i.e., gate-source offset driving) to cut leakage during non-selected periods, increase the cell read current. The resulting degraded static noise margin (SNM) is compensated for by increased conductance in cross-coupled MOSFETs caused by the raised supply (V_{DH}), which is generated by the charge pump or the voltage-down converter of the I/O power supply. Furthermore, the V_{DH} maintains the signal charge Q_S and drivability of cross-

Fig. 7-14   Static noise margin (SNM) and cell read current (*Icell*) of raised power supply cell at 1-V V_{DD} for low V_T (a) and high V_T (b) [4, 36].

coupled MOSFETs despite the high V_T and large variations in V_T and V_T-mismatch (δV_T). The δV_T not only makes the cell read current (I_{cell}) smaller, but also makes SNM imbalanced and thus narrower in its so-called butterfly curve for the worst combination of V_T in the figure, as exemplified by a δV_T of 0.1 V in Figure 7-13(b). Even with a small boost of $\delta V_D = 0.1$ V for $\delta V_T = 0.1$ V both I_{cell} and SNM are greatly increased for both a high-speed design of $V_T = 0.49$ V and a low-power design of $V_T = 0.71$ V (see Figure 7-14) [4, 36]. Thus, the V_{DH} scheme is likely to make 1-V operation possible even under the usual design conditions of a 100-mV SNM and 20-μA cell current, while the conventional V_{DD} scheme (i.e., no raised power supply and fixed high-V_T transfer MOSFETs) makes 1-V operation at $V_T = 0.71$ V impossible. Power dissipation with this V_{DH} scheme is reduced because the data-line (DL) voltage can be maintained low (i.e., V_{DD}) to reduce the power of DL and other column relevant circuits that are major sources of SRAM power. Thus, even if a high voltage is applied to the word line and cell, power dissipation of an SRAM will not be substantially increased.

7.7 APPLICATIONS TO PERIPHERAL CIRCUITS

7.7.1 DRAM Peripheral Circuits

SSI was applied to iterative circuit blocks to reduce leakage current, first in the standby mode and then in the active mode with advanced SSIs.

Standby-current Reductions in 256-Mb DRAMs: SSI dramatically reduced the data retention current with refresh operations that was dominated by the word-driver block. Figure 7-15 shows low-V_T p-MOS SSI (Q_S) shared with the n word-driver block in a 256-Mb DRAM [39]. This is an example of A1 in Table 7-1 or Figure 7-4, although a raised supply (V_{DH}) necessary for word-bootstrapping is used. In the standby mode, off-SSI enables the voltage (V_{DL}) of the common source line to drop by δ from V_{DH} as a result of the total sub-threshold current flow of ni. As each p-MOS driver (Q_P) is self-reverse-biased, the total current eventually decreases. Hence, the V_{DH} is well regulated despite the use of the on-chip charge pump. In the active mode, the selected word line is driven after V_{DL} becomes V_{DH} by turning on Q_S. Here, the channel width of Q_S can be reduced to several times that of Q_P without a

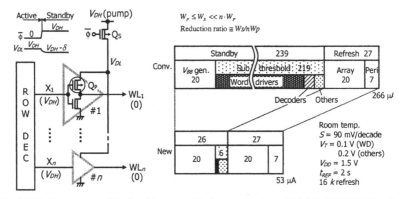

Fig. 7-15   Gate-source (V_{GS}) self-reverse biasing applied to a 256-Mb DRAM (a) and retention current with refreshes [39]. W_S and W_P denote the respective channel widths of Q_S and Q_P.

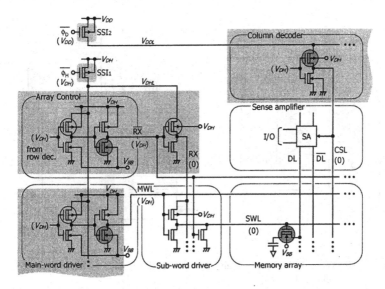

Fig. 7-16 Various leakage-reduction schemes applied to array associated
circuitry in a 256-Mb SDRAM [22]. ( ): voltages in the standby mode.

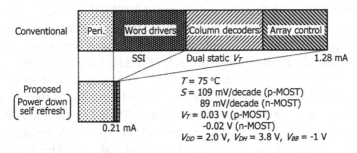

Fig. 7-17 Leakage in the standby mode of 256-Mb SDRAM [22].
The peripheral circuits component is from peripheral MOSTs
without substrate bias.

speed penalty, as previously discussed. A resulting δ as small as 254 mV
reduced the standby sub-threshold current of the word-driver and decoder
blocks to 1.5×10^{-3} for $n = 256$ and $W(Q_S)/W(Q_P) = 5$, enabling to reduce the
sub-threshold current of the chip to 3% (from 219 to 6 µA). The total data
retention current was thus reduced to 53 µA. The small δ enabled a fast
recovery time to the active mode of 1 ns.

Figure 7-16 shows another 256-Mb DRAM [22] with a hierarchical
word-line architecture [2]. The SSI and multi-static high V_T utilizing well
biases have been combined. Here, /MWL and SWL are the main word line
and sub-word line, and CSL is the column select line for the multi-divided
data-line architecture [2]. The circled MOSFETs in the figure are in the sub-
threshold region during the standby mode. Here, SSI_1 was only applied to
the p-MOSFETs (open circles) in the inverter chain with an output of RX in

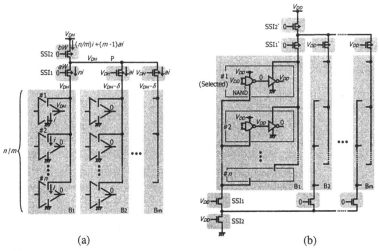

Fig. 7-18 Word driver block (a) and decoder block (b) with the hierarchical SSI
scheme [13, 14].

the array control circuit, which corresponds to Figure 7-7(a). This was
because p-MOSFETs have a larger total channel width and a larger S-factor
(see Figure 7-17) due to their buried-channel MOSFET structure. Since these
inverters operate in series in the time domain, this connection does not cause
a speed penalty, as previously explained. SSI_1 was also applied to the main-
word driver block. Furthermore, the n-MOSFETs (shaded circles) and p-
MOSFETs (shaded circles) in the column-decoder block had a higher V_T due
to the respective well bias V_{BB} and V_{DH}. SSI_2 further reduced the leakage in
the column decoder block. By combining both SSI and multi-static high-V_T
schemes, the total sub-threshold current in the power-down/self-refresh
mode was reduced to one-sixth, as Figure 7-17 shows. The current can
further be reduced by applying a multi-static V_T scheme to the peripheral
circuits.

Active-current Reduction of 16-Gb DRAM: Hierarchical SSI schemes (i.e.,
Table 7-2(d)) and a power switch were effective in reducing the active
current of a hypothetical 1-V 16-Gb DRAM [13, 14]. Figures 7-18(a) shows
application to the word-driver block [13, 14], which is divided into m sub-
blocks with n/m word drivers each. An SSI_1 is connected to the common
source line of p-MOSFET word drivers to select the sub-block. In the active
mode, while turning on the selected SSI1 and SSI2, the leakage in each non-
selected sub-block is confined to the SSI_1 small constant current since the
SSI_1 channel width is much narrower than the total channel width of driver
MOSFETs in the corresponding sub-block, as previously discussed. In the
standby mode, the total leakage confined by SSI_1s (i.e., *mai* in Table7-2(c))
is further confined to the SSI_2 small constant current since the SSI_2 channel

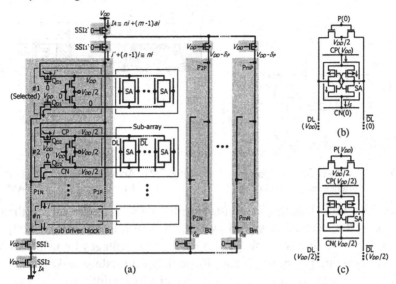

Fig. 7-19 Sense amp (SA) driver block (a) with a symmetric hierarchical SSI
scheme in the active mode, and selected SA-relevant circuits (b) and
non-selected SA-relevant circuits (c)[13, 14]. P; Precharge.

width is much narrower than the total channel width of SSI_1s. Figure 7-18(b)
shows application to the decoder block, which is divided into m sub-blocks
(B_1-B_m) with n drivers each. One decoder unit consists of an n-MOSFET
NAND gate with address inputs and a CMOS inverter. The hierarchical SSI
scheme is applied to the n-MOSFET common source of the NAND gate
block, and to the p-MOSFET common source of the CMOS inverter block.
Leakage current in active and standby modes is reduced in the same manner
as in the word driver block.

Figure 7-19(a) shows application to the sense-amp driver block [13, 14],
which is divided into m sub-blocks (B_1-B_m) with n drivers each. In the active
mode, the selected driver in the selected sub-block, e.g., driver #1 in sub-
block B_1, drives the sense amps (SAs) in the selected sub-array with Q_{D1} and
Q_{D1}'. This is done by turning on SSI_2 and SSI_2', and only SSI_1 and $SSI_1'$of
selected sub-block B_1, while all SSI_1s and SSI_1's of non-selected sub-blocks
remain off. After the signals developed on the half-V_{DD} data lines have been
amplified to V_{DD} or 0 by the selected SAs, accumulated leakage (i'), which is
the sum of leakage (i_S) flowing into each SA (Figure 7-19(b)), flows into Q_{D1}
and Q_{D1}'. The accumulated leakage is proportional to the total channel width
of leakage-relevant MOSFETs in SAs, which is comparable to the channel
width of Q_{D1} and Q_{D1}' to quickly drive the SAs. For other drivers in the
selected block, all SAs remain off because all data lines and all drive lines
(CPs and CNs) are at $V_{DD}/2$ (Figure 7-19(c)). However, leakage (i), which is
proportional to the channel width of Q_D or Q_D', develops at each non-
selected (OFF) driver p-MOSFET and n-MOSFET (e.g., Q_{D2} and Q_{D2}').

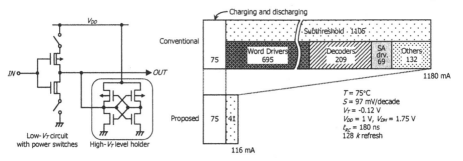

Fig. 7-20 Low-V_T circuit with power   Fig. 7-21 Active current reduction in the
switches and a level holder [14].   hypothetical 16-Gb DRAM [14].

Hence, the total leakage in non-selected drivers is much larger than the leakage in the selected driver. However, for non-selected blocks, SSI_1s and SSI_1's enable to develop voltage drops or raised voltages by using δ_P or δ_N on corresponding power lines (e.g., P_{2N} and P_{2P}.) to reduce leakage. The total leakage is $(m-1)$ ai, assuming that the channel width ratio of an SSI_1 MOSFET to Q_D is a. Note that Q_{D1} and Q_{D1}'are common to plural SAs in this example. However, they can be distributed to each SA without an area penalty, as can be seen in Figure 7-22.

Figure 7-20 shows a high-V_T power switch applied to an internal low-V_T circuit with a high-V_T level holder [14]. It is useful for various control logic circuits in peripheral circuits to maintain the output level, since a floating intermediate level may cause leakage in succeeding circuits. The input of a low-V_T circuit is evaluated with this scheme, and then the evaluated output is maintained at the small level holder. After that, the power switches are turned off to stop leakage in the low-V_T circuit, while preventing output from unnecessary discharge. Thus, the switches can be turned on quickly at the necessary timing to prepare the next evaluation, ensuring fast random logic operation despite the large area and large voltage swing to control the power switches.

In the conventional design, the simulated active current in a 16-Gb DRAM was as large as 1.18 A, (see Figure 7-21). The DC sub-threshold leakage was as large as 1.105 A, while the AC capacitive current was as small as 75 mA at a cycle time of 180 ns. Major leakage current came from iterative circuit blocks, such as the word driver block, decoder blocks (X and Y), and sense-amp driver blocks. Note that depletion MOSFETs, e.g., a - 0.12-V V_T for n-MOSFETs, were responsible for such large leakage. The circuits we described reduced the active current to 116 mA.

Sleep-mode Current Reduction in 0.09-μm 16-Mb e-DRAM: A variety of SSI schemes and compensation circuits for V_T-variations has enabled a 16-Mb e-DRAM to operate at a record-setting low voltage of 0.6 V using a 0.195-μm² trench capacitor ($C_S = 40$ fF) 1-T cell [1]. The total operating power at a 0.6-

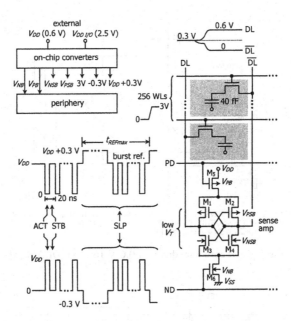

Fig. 7-22   Cell relevant circuits of 0.6-V 16-Mb e-DRAM [1].
STB: standby mode, ACT: active mode, and SLP: sleep mode.

V 20-ns row cycle was only 39 mW, and the standby and sleep-mode currents at 0.9 V and 105°C were as low as 328 and 34 μA. Details on the circuits are explained in what follows.

Figure 7-22 shows the cell-relevant circuits. An excessively raised word voltage of 3 V is probably needed for high-speed charging of a large C_S rather than for a full-V_{DD} write. A low V_T of 0.2 V is used for SA MOSFETs (M1-M4) to enable high-speed half-V_{DD} sensing, while a normal V_T of 0.3 V is used for SA-driver MOSFETs (M5, M6) to reduce sub-threshold current in standby mode. The substrate biases of SA and driver MOSFETs are independently controlled because their different V_T implants cause different temperature dependencies for V_T. Gate-source offset driving in the sleep mode of n-MOSFETs and p-MOSFETs (above V_{DD} and below V_{SS} by 0.3 V) reduces sub-threshold current. The RAM data is retained by periodically exiting the sleep mode and performing burst refresh cycles at 20 ns, as we can see from the figure. The refresh scheme minimizes the pump currents of the converters for -0.3 V and V_{DD} +0.3 V, enabling a simple design to be used for the converters. Figure 7-23 depicts a sense-amp driver consisting of an inverter chain that shares SSIs, as shown in Figure 7-7(a). Figure 7-24 shows row circuits composed of multiple iterative circuit blocks such as a NAND decoder block, an inverter block, a level shifter block, and a word-driver block. Each block has its own SSI MOSFET. For example, in the sleep mode, the leakage from each block is reduced by the respective SSI because each circuit in the block and SSI are off. In the active mode, all SSIs

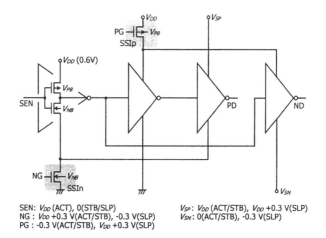

SEN: V_{DD} (ACT), 0(STB/SLP)
NG : V_{DD} +0.3 V(ACT/STB), -0.3 V(SLP)
PG : -0.3 V(ACT/STB), V_{DD} +0.3 V(SLP)

V_{SP}: V_{DD} (ACT/STB), V_{DD} +0.3 V(SLP)
V_{SN}: 0(ACT/STB), -0.3 V(SLP)

Fig. 7-23   Sense-amp driver of 0.6-V 16-Mb e-DRAM [1].

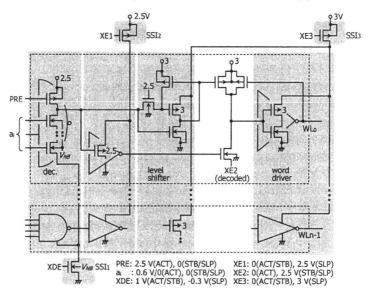

XDE: 1 V(ACT/STB), -0.3 V(SLP)

PRE: 2.5 V(ACT), 0(STB/SLP)   XE1: 0(ACT/STB), 2.5 V(SLP)
a_i : 0.6 V/0(ACT), 0(STB/SLP)   XE2: 0(ACT), 2.5 V(STB/SLP)
XDE: 1 V(ACT/STB), -0.3 V(SLP)   XE3: 0(ACT/STB), 3 V(SLP)

Fig. 7-24   Row circuits of 0.6-V 16-Mb e-DRAM [1].

are turned on so that the selected word line is activated by the corresponding row circuits. Only one circuit in each block is activated in the active mode, and thus leakage in the sleep mode is reduced without speed and area penalties, as previously described.

The substrate biases of SAs and other periphery circuits must statically be controlled according to variations in process, temperature, and V_{DD} to suppress variations in sub-threshold-current as well as speed. An n-MOS body-bias generator [1], shown in Figure 7-25(a), monitors the current through reference MOSFET M_1 with V_{GS} approximately equal to V_T ($V_{DD}/2$). The current through the MOSFET (I_{DS}) is a good indicator of both

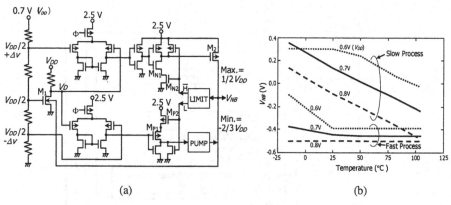

Fig. 7-25 Body-bias generator for n-MOSTs (a) and generated V_{NB} (b) [1].

MOSFET-OFF current and the switching speed of peripheral circuits. The drain voltage (V_D) is compared to $V_{DD}/2 + \Delta v$ and $V_{DD}/2 - \Delta v$ using two OP-amps to determine whether the body bias (V_{NB}) should be increased or decreased. When the V_T of peripheral circuits is low due to fast process conditions or high temperatures to the extent of $V_D < V_{DD}/2 - \Delta v$, the lower OP-amp senses the reduced V_D so that PUMP (i.e., on-chip charge pump [2]) starts the built-in ring oscillator to oscillate if M_{P2} is on. Thus, V_{NB} starts to decrease so that the V_T is increased to compensate. The oscillations continue until the resulting deep V_{NB} increases the V_T to a point where the OP-amp turns M_{P1} off with $V_D > V_{DD}/2 - \Delta v$. This is true as long as the deep V_{NB} does not exceed the acceptable lower limit for V_{NB}. Once V_{NB} exceeds the lower limit for V_{NB} on the way to the deep V_{NB}, LIMIT detects the level of the lower limit for V_{NB} and turns M_{P2} off with the "H" output to inhibit oscillation. The upper OP-amp works in the same manner, comparing V_D to $V_{DD}/2 + \Delta v$. However, when V_T is high due to slow process conditions or low temperatures, the lower OP-amp disables PUMP while the upper OP-amp discharges the M_2 gate for driving the body. Thus, the V_T of n-MOSFETs is reduced and thus compensated for. Another feature of the generator is that as V_{DD} increases, body bias V_{NB} becomes negative to raise V_T and reduce standby power with reduced sub-threshold current. The positive body bias is limited to a maximum of $V_{DD}/2$ to limit the leakage into the body-to-source/drain junctions, and the negative body bias is limited to $-2/3V_{DD}$ to avoid over-stressing the MOSFETs. Figure 7-25(b) plots V_{NB} versus temperature as a function of V_{DD}. The p-MOSFET body bias generator is a complementary version of the n-MOSFET generator. It has been reported that negative body bias reduced sub-threshold currents by 75 % under fast process conditions, and positive body bias improved the speed by 63 % under slow process conditions.

The output level of each internal voltage in Figure 7-22 is well regulated if the loads that on-chip pumps must drive are kept light. However, possible

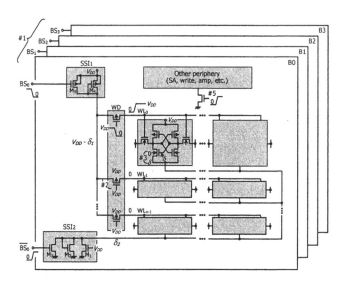

Fig. 7-26   Multi-bank architecture applied to 0.13-μm 1.2-V 1-Mb e-SRAM module [41].

problems [2] are detrimental effects due to parasitic bipolar transistors that develop more easily at forward bias [40].

7.7.2 SRAM Peripheral Circuits

Figure 7-26 illustrates a 0.13-μm 1.2-V 300-MHz 1-Mb e-SRAM module [41]. In the active mode, a four-bank architecture with only one-bank activation (#1 in the figure) by turning on M_3 and M_5 confines the active circuitry to one-fourth, and thus reduces the AC power of control signals to one-fourth. This also reduces leakage in inactive banks, if SSI is applied to the word-driver block and cell array. SSI_1 reduces the leakage in each off-p-MOSFET in the word driver block [4] by gate-source self-reverse biasing (#2). SSI_1 allows a small voltage drop, δ_l, in the small capacitive power line of the word-driver block, enabling a fast recovery time of 0.3 ns. SSI_2 causes δ_2 on the common cell-source line as a result of accumulated leakage from many cells. Thus, the body effect by δ_2 (#3) increases the V_T of the cross-coupled off-n-MOSFET in each cell so that leakage is reduced. Here, the diode (M_2) clamps the source voltage, so signal charge Q_S is not reduced by excessive δ_2. However, it has drawbacks: A large necessary δ_2 of about 0.4 V due to poor reduction efficiency of source-driving (B2 in Figure 7-4), and large source-line capacitance result in a slow recovery time of about 3 ns. In addition, the signal charge of the non-selected cell is reduced by δ_2. In the sleep mode, peripheral circuits such as SAs and write amps are turned off with the power switch off, resulting in a slow recovery time of 3 ns. In the standby mode, all banks are off, causing further reduced leakage.

The results indicated that leakage current was reduced by 25% in the high-speed (300 MHz) active mode with only SSI_1 turned on, and by 67% in the slow-speed (100 MHz) active mode with both SSI_1 and SSI_2 turned on. In the sleep mode, leakage in peripheral circuits was reduced by 95%. Even so, a 1.2-V V_{DD} is still high, and Q_S reduced by δ_2 in the standby mode would pose a stability problem at a lower V_{DD}.

7.7.3 Control Logic Circuits

The total channel width of MOSFETs involving shubthreshold currents in high-speed control logic circuits is relatively small, as previously explained. Thus, few reduction schemes have been proposed thus far. If needed, however, high-speed reduction schemes such as various SSI schemes, a power switch with a level holder, and multi-static V_T utilizing internal voltages can successfully be applied to circuits. These reduction schemes may also be used in the circuits of logic LSIs because they have similar circuit configurations. In fact, SSI enabling the leakage of an adder in logic LSIs to be controlled within one active cycle has been reported [42].

7.8 FUTURE PROSPECTS

A dual approach for V_{DD}, V_T, and t_{OX} [3] will eventually be needed to meet different cell and peripheral circuit requirements. For RAM cells, leakage and cell stability are major concerns. To reduce leakage, the necessary V_T must be high, and gradually be increased with memory capacity to preserve the refresh busy rate [2] for DRAMs and the chip retention current for SRAMs, as previously discussed. To achieve greater stability, such high V_T unavoidably necessitates a high V_{DD} for a large signal charge, and thus a thick t_{OX} for a small gate tunneling current. In contrast, for the peripheral circuits, low power and high speed are major concerns, as they are for logic LSIs. Thus, a low and scalable V_T, a low V_{DD}, and a thin t_{OX} are necessary. Therefore, for example, the difference in V_T between RAM cells and peripheral circuits will increase in the future. Even so, since the difference should be reduced for simple designs, leakage issue will continue to be important.

New gate insulators to reduce the gate tunneling current need to be developed soon. In addition, the development of new devices and circuits to suppress leakage variations as well as speed variations caused by V_T-variations will be real challenges to future RAM designs. Increased chip-to-chip leakage by inter-die V_T-variations, however, could be compensated for by further improvements of existing circuits using internal supply voltages, as discussed earlier. Intra-die V_T-variations degrade the voltage margins of

SRAM cells, DRAM sense amps, and peripheral circuits unacceptably, making reliable RAM designs almost impossible for the near future. Hence, new MOSFETs with small V_T-variations are necessary. Although in the long run, low-temperature CMOS operation might be a possibility, the fully-depleted (FD) SOI may be a strong candidate. For example, if a new FD-SOI device, called a dynamic-double-gate SOI (D2G-SOI) [43-45], is used, many advantages that make them suitable for sub-volt operations are expected. These are not only smaller V_T-variations and thus smaller V_T-mismatch variations (Figure 7-2), but also a larger signal charge with additional capacitances, a smaller necessary signal charge with a soft-error immune structure, a lower necessary V_T with a reduced S-factor, a wider static noise margin with increased drain current, and a reduced temperature dependence of V_T [46].

7.9 CONCLUSION

This chapter presented mainly challenges and trends in low-voltage RAMs using nanoscale transistors in terms of sub-threshold leakage current in RAM cells and peripheral circuits. After comparing the many schemes that have been proposed to date to reduce leakage, the switched-source impedance scheme was discussed in detail in terms of speed, area penalty, and active-leakage reduction capabilities. We concluded it to be the most suitable to reduce leakage in RAMs in active and standby modes. Then, various applications of the scheme to DRAMs and SRAMs were discussed. Based on these considerations, future prospects were considered with emphasis on needs for a dual approach for V_{DD}, V_T, and t_{OX}, precise controls of internal voltages to reduce variations in leakage as well as speed caused by V_T variations, and new MOSFETs such as fully-depleted SOIs with small V_T-variations.

REFERENCES

[1] K. Hardee et al., "A 0.6V 205MHz 19.5ns t_{RC} 16Mb Embedded DRAM," *2004 ISSCC Dig. Tech. Papers*, pp. 494-495, Feb. 2004.

[2] K. Itoh, *VLSI Memory Chip Design*, Springer-Verlag, NY, 2001.

[3] Y. Nakagome et al., "Review and prospects of low-voltage RAM circuits," *IBM J. R & D*, vol. 47, no. 5/6, pp. 525-552, Sep. /Nov. 2003.

[4] K. Itoh et al., "Reviews and Prospects of Low-voltage Embedded RAMs," *CICC2004 Dig. Tech. Papers*, pp. 339-344, Oct. 2004.

[5] M. Aoki et al., "A 1.5V DRAM for Battery-Based Applications," *ISSCC Dig. Tech. Papers*, pp. 238–239, Feb. 1989.

[6] Y. Nakagome et al., "A 1.5-V Circuit Technology for 64Mb DRAMs," *Symp. VLSI Circuits Dig. Tech. Papers*, pp. 17–18, June 1990.

[7] K. Itoh, "Reviews and Prospects of Deep Sub-Micron DRAM Technology," *SSDM Ext. Abst.*, pp. 468–471, August 1991.

[8] J. Etoh et al., "Large Scale Integrated Circuit for Low Voltage Operation," *U.S. Patent 5, 297, 097*, March 1994.

[9] Y. Nakagome et al., "Sub-1-V Swing Bus Architecture for Future Low-Power ULSIs," *Symp. VLSI Circuits Dig. Tech. Papers*, pp. 82–83, June 1992.

[10] T. Kawahara et al., "Sub-threshold Current Reduction for Decoded-Driver by Self-Reverse Biasing," *IEEE J. Solid-State Circuits*, Vol. 28, No. 11, pp. 1136–1144, Nov. 1993.

[11] M. Horiguchi et al., "Switched-Source-Impedance CMOS Circuit for Low Standby Sub-threshold Current Giga-Scale LSI's," *IEEE J. Solid-State Circuits*, Vol. 28, pp. 1131–1135, No. 11, Nov. 1993.

[12] D. Takashima et al., "Standby/Active Mode Logic for Sub-1-V Operating ULSI Memory," *IEEE J. Solid-State Circuits*, Vol. 29, No. 4, pp. 441–447, April 1994.

[13] T. Sakata et al., "Sub-threshold-current Reduction Circuits for Multi-gigabit DRAMs," *Symp. VLSI Circuits Dig. Tech. Papers*, pp. 45-46, May 1993.

[14] T. Sakata et al., "Sub-threshold-Current Reduction Circuits for Multi-Gigabit DRAM's," *IEEE J. Solid-State Circuits*, Vol. 29, No. 7, pp. 761–769, July 1994.

[15] S. Mutoh et al., "1-V Power Supply High-Speed Digital Circuit Technology with Multithreshold-Voltage CMOS," *IEEE J. Solid-State Circuits*, Vol. 30, No. 8, pp. 847–854, August 1995.

[16] K. Nii et al., "A 90 nm Low Power 32K-Byte Embedded SRAM with Gate Leakage Suppression Circuit for Mobile Applications," *Symp. VLSI Circuits Dig.*, pp. 247-150, June 2003.

[17] H. R. Huff and D. C. Gilmer (Eds.), *High Dielectric Constant Materials-VLSI MOSFET Applications*, Springer-Verlag, 2004.

[18] T. Inukai and T. Hiramoto, "Suppression of Stand-By Tunnel Current in Ultra-Thin Gate Oxide MOSFETs by Dual Oxide Thickness MTCMOS," *SSDM Ext. Abst.*, pp. 264–265, August 1999.

[19] P. M. Carter and B. R. Wilkins, "Influences on Soft Error Rates in Static RAM's," *IEEE J. Solid-State Circuits*, Vol. sc-22, No.3, pp. 430-436, June 1987.

[20] S-M Jung et al., "Soft Error Immune 0.46 μm^2 SRAM Cell with MIM Node Capacitor by 65 nm CMOS Technology for Ultra High Speed SRAM," *IEDM Tech. Dig.*, pp. 289-292, Dec. 2003.

[21] K. Itoh, "Reviews and Prospects of Low-Power Memory Circuits" (invited), *Low-Power CMOS Design*, A. Chandrakasan and R. Brodersen, Eds., Wiley–IEEE Press, NJ, pp. 313–317, 1998.

[22] M. Hasegawa et al., "A 256Mb SDRAM with Sub-threshold Leakage Current Suppression," *ISSCC Dig. Tech. Papers*, pp. 80–81, Feb. 1998.

[23] K. Itoh and H. Mizuno, "Low-voltage Embedded-RAM Technology: Present and Future," *Proc. 11th IFIP Int'l Conf. VLSI*, pp.393-398, Dec. 2001.

[24] S. Heo and K. Asanovic, "Leakage-Biased Domino Circuits for Dynamic Fine-Grain Leakage Reduction," *Symp. VLSI Circuits Dig. Tech. Papers*, pp. 316–319, June 2002.

[25] S. Narendra et al., "Scaling of Stack Effect and its Application for Leakage Reduction," *Proc. ISLPED*, pp. 195–199, August 2001.

[26] C. Akrout et al., "A 480-MHz RISC Microprocessor in a 0.12-μm L_{eff} CMOS Technology with Copper Interconnects," IEEE *J. Solid-State Circuits*, Vol. 33, No. 11, 1609–1616, Nov. 1998.

[27] H. Morimura and N. Shibata, "A 1-V 1-Mb SRAM for Portable Equipment," *Proc. ISLPED*, pp. 61–66, August 1996.

[28] K. Itoh et al., "A Single 5V 64K Dynamic RAM," *ISSCC Dig. Tech. Papers*, pp. 228–229, Feb. 1980.

[29] I. Fukushi et al., "A Low-Power SRAM Using Improved Charge Transfer Sense Amplifiers and a Dual-Vth CMOS Circuit Scheme," *Symp. VLSI Circuits Dig. Tech. Papers*, pp. 142–145, June 1998.

[30] A. Keshavarzi et al., "Effectiveness of Reverse Body Bias for Leakage Control in Scaled Dual Vt CMOS ICs," *Proc. ISLPED, pp. 207–212*, August 2001.

[31] H. Mizuno et al., "A 18-μA Standby Current 1.8-V 200-MHz Microprocessor with Self-Substrate-Biased Data-Retention Mode," *IEEE J. Solid-State Circuits,* Vol. 34, No. 11, pp 1492–1500, Nov. 1999.

[32] T. Sakata et al., "Two-Dimensional Power-Line Selection Scheme for Low Sub-threshold-Current Multi-Gigabit DRAMs," *Proc. ESSCIRC*, pp. 131–134, Sept. 1993,.

[33] K. Seta et al., "50% Active-Power Saving Without Speed Degradation Using Standby Power Reduction (SPR) Circuit," *ISSCC Dig. Tech. Papers*, pp. 318–319, Feb. 1995.

[34] K. Kumagai et al., "A Novel Powering-Down Scheme for Low Vt CMOS Circuits," *Symp. VLSI Circuits Dig. Tec. Papers*, pp. 44–45, June 1998,

[35] M. Mizuno et al., "Elastic-Vt CMOS Circuits for Multiple On-Chip Power Control," *ISSCC Dig. Tech. Papers*, pp. 300–301, Feb. 1996.

[36] K. Ioth et al., "Low-voltage Embedded RAMs-Current Status and Future Trends," E. Macii et al. (Eds.): *PATMOS2004, LNCS3254*, pp. 3-15, 2004, ©Springer-Verlag Berlin Heidelberg 2004.

[37] K. Osada et al., "16.7fA/Cell Tunnel-Leakage-Suppressed 16-Mbit SRAM Based on Electric-Field-Relaxed Scheme and Alternate ECC for Handling Cosmic-Ray-Induced Multi-Errors," *ISSCC Dig. Tech. Papers*, pp. 302–303, Feb. 2003.

[38] K. Itoh et al., "A Deep Sub-V, Single Power-Supply SRAM Cell with Multi-Vt, Boosted Storage Node and Dynamic Load," *Symp. VLSI Circuits Dig. Tech. Papers*, pp. 132–133, June 1996.

[39] G. Kitsukawa et al., "256-Mb DRAM Circuit Technologies for File Applications," *IEEE J. Solid-State Circuits,* Vol. 28, No. 11, pp. 1105–1113, Nov. 1993.

[40] G. Ono et al., "Temperature Referenced Supply Voltage and Forward-body-bias Control (TSFC) Architecture for Minimum Power Consumption," *ESSCIRC Dig. Tech. Papers*, pp. 391-394, Sept. 2004.

[41] M. Yamaoka et al., "A 300MHz 25µA/Mb Leakage On-Chip SRAM Module Featuring Process-Variation Immunity and Low-Leakage-Active Mode for Mobile-Phone Application Processor," *ISSCC Dig. Tech. Papers*, pp. 494-495, Feb. 2004.

[42] T. Miyazaki et al., "Observation of one-fifth-of-a-clock wake-up time of power-gated circuit," *CICC Dig. Tech. Papers*, pp. 87-90, Oct. 2004.

[43] M. Yamaoka et al., "Dynamic-Vt, Dual-Power-Supply SRAM Cell using D2G-SOI for Low-Power SoC Application," *Int'l SOI Conf. Dig. Tech. Papers,* pp. 109-111, Oct. 2004.

[44] R. Tsuchiya et al., "Silicon on Thin BOX: A New Paradigm of The CMOSFET for Low- Power and High-Performance Application Featuring Wide-Range Back-Bias Control," *IEDM Dig. Tech. Papers*, Dec. 2004.

[45] M. Yamaoka et al., "Low Power SRAM Menu for SOC Application Using Yin-Yang-Feedback Memory Cell Technology," *Symp. VLSI Circuits Dig. Tech. Papers*, pp. 288-291, June 2004.

[46] G. Groeseneken et al., "Temperature Dependence of Threshold Voltage in Thin-Film SOI MOSFET's," *IEEE Electron Device Letters*, Vol. 11, No. 8, pp. 329-331, August 1990.

Chapter 8

ACTIVE LEAKAGE REDUCTION AND MULTI-PERFORMANCE DEVICES

Siva Narendra[§], James Tschanz[¶], Shekar Borkar[¶], and Vivek De[¶]
§*Tyfone, Inc., USA and* ¶*Intel Corp., USA*

8.1 INTRODUCTION

As leakage current components increase they not only affect the standby power consumption, but also active power consumption. Under the current scaling trends leakage current is expected to be comparable to switching power. Under this scenario is it is not sufficient to deal with leakage power only during standby mode. Reducing leakage power in standby mode is often easier, since the performance of the CMOS circuit is not relevant in standby mode. In active mode however performance of the active CMOS circuit cannot be compromised to reduce leakage power. In this chapter we describe two solutions for reducing the active leakage of CMOS circuits: dynamic leakage reduction techniques, and multi-performance device insertion.

Dynamic leakage reduction techniques are applicable in high performance CMOS systems where not all units are active at the same time, as shown in Figure 8-1 [1]. In present designs idle units utilize clock gating to reduce switching power, but these units continue to consume leakage power. Traditional standby leakage power reduction techniques can now be applied on a dynamic basis to these idle units. As illustrated in the figure, dynamic sleep transistor and dynamic body bias techniques can be used to reduce leakage power of idle units, thereby reducing the overall active leakage power. As described in Chapter 2 stack effect could also be used to reduce active leakage power.

The applicability of a standby leakage technique for active leakage reduction will depend on the time constant necessary to achieve leakage reduction, time required to deactivate the leakage reduction mode, and energy overhead in switching in and out of the leakage reduction mode. In this chapter active leakage reduction using dynamic body bias and dynamic sleep transistor in conjunction with clock gating is used to show the reduction in switching and leakage powers components.

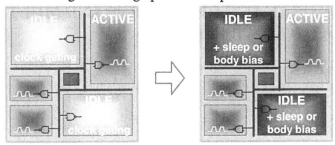

Figure 8-1. Illustration of system level active leakage reduction.

Another method to save active leakage and standby leakage at the same time is to use lower leakage devices in paths that are not performance critical. Power-performance trade-off can be achieved through optimal use of multi-performance devices. One such case is the use of dual or multiple threshold voltage process technology. Multi-performance devices can also be achieved through different channel lengths or by using stacked devices for achieving power-performance trade-off. Use of multi-performance devices through dual threshold voltage process is discussed in this chapter as well.

8.2 STANDBY TECHNIQUES FOR ACTIVE LEAKAGE REDUCTION

Clock gating is used in high performance microprocessors to reduce average active power and energy consumptions [2]. Disabling the clock to idle functional units saves power by preventing wasteful switching power dissipation in the local clock distribution network and sequentials during the idle period. However, with technology scaling, leakage power of idle units is becoming a large fraction of the total chip power. As a result, the overall power savings achievable by clock gating alone is diminishing. Using dynamic power gating and body bias techniques in conjunction with clock gating can help to control the active leakage power of an ALU in a 32-bit integer execution core [3]. Performance impacts, area overheads, active leakage and total power reductions achievable by these techniques are measured on the prototype chip (Figure 8-2) implemented in a 130nm dual-V_t CMOS technology.

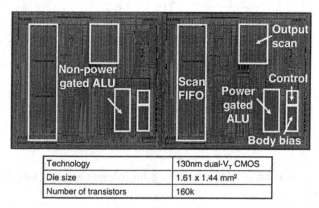

Technology	130nm dual-V_T CMOS
Die size	1.61 x 1.44 mm²
Number of transistors	160k

Figure 8-2. Dynamic power gating (sleep transistors) and body bias case study

The 32-bit, 2-phase domino, Han-Carlson adder design contains NMOS and PMOS power gating transistors, inserted between the virtual and real supply grids on chip (Figure 8-3). The power gating transistors are ON during active mode and are turned OFF during the idle phase along with the local clock. To further improve on-resistance and leakage reduction capabilities of the power gating transistors we use combinations of (1) gate overdrive during "active" and underdrive during "idle", and (2) forward body bias (FBB) during "active" and zero body bias (ZBB) during "idle".

Low-V_t devices are used for the power gating transistors to minimize performance and area impacts. The power gating transistors are distributed uniformly across the adder layout to prevent undesirable current crowding in the power grids. M4 and M5 metal levels in the traditional power grid are re-connected through the power gating transistors instead of M4/M5 vias. The virtual supply transition waveform, following power gating transistor turn-off, can be measured directly using an on-chip 8-level (3-bit) A/D converter that uses inverters of varying pre-characterized trip points. Dynamic body bias for the adder core, to apply FBB during "active" and ZBB during "idle", is also implemented.

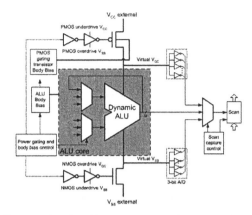

Figure 8-3. Block diagram of ALU with power gating (sleep) transistors, body bias, and control circuitry.

High frequency testing is accomplished using a scan methodology for the ALU input vectors and at-speed output capture. Measurement of the virtual supply transition waveform, following power gating transistor turn-off, is done by capturing outputs of the two 3-bit A/D converters after a preset number of clock cycles. Separate supply pins are provided for the buffers driving the power gating transistors so that switching and leakage power of the adder can be measured independently of the energy required to switch the power gating transistors between active and idle.

The adder operation frequency, without power gating transistors, ranges from 3.3 GHz (1.1 V) to 4.3 GHz (1.4 V) at 75°C with FBB applied to the adder core. Using PMOS power gating transistors degrades frequency by 2.3%, with an associated area overhead of 11%. Idle leakage power is 13X smaller at 75°C (Figure 8-4).

Gate oxide leakage through the MOS decoupling capacitors (decap) on the real supply grid cannot be reduced by power gating transistor. The adder leakage power reduction is 37X when the decap leakage is excluded. When 200 mV gate overdrive and underdrive are used, the frequency loss is 1.8%, but leakage power savings remain largely unchanged because the decap leakage is significant.

Using 450 mV FBB for the PMOS power gating transistors reduces frequency impact from 2.3% to 1.8%. In contrast to the power gating transistor technique, using dynamic body biasing for PMOS devices in the adder core reduces leakage power by 1.8X. The area overhead for body bias generators and bias grid routing is 2%. While body biasing is useful for controlling sub-threshold leakage only, power gating transistors help reduce gate oxide, junction and sub-threshold leakage components of the leakage power.

However, logic state of the adder is lost during the "idle" period when power gating transistors are turned off, but is preserved when dynamic body biasing is used.

Reference case: adder without power gating transistors, 450mV FBB (1.28V, 75°C, 4GHz)	Frequency change	Leakage reduction	Leakage reduction including decap oxide leakage	Area overhead
PMOS power gating transistor	-2.3%	37X	13.3X	11%
PMOS power gating transistor with 200mV overdrive/underdrive	-1.8%	44X	13.8X	12%
PMOS power gating transistor with 450mV FBB/ 500mV RBB	-1.8%	64X	15.5X	12%
Adder with PMOS FBB during active mode & ZBB during idle	0%	1.9X	1.8X	2%

Figure 8-4. Adder frequency and leakage with PMOS power gating (sleep) transistor, compared to no power gating transistor.

Overall performance impact of these dynamic leakage control techniques is dictated by the time required to switch the power gating transistor gate or the body node voltage during "idle" to "active" transition. Measurements and simulations show that this time is 1 clock cycle for power gating transistors and 3-4 cycles for body bias, compared to less than a cycle for clock gating. For any of these dynamic leakage control techniques to achieve reduction in overall power, the leakage energy saved during the "idle" period must be larger than any energy overhead incurred during transitions between "idle" and "active" modes.

For power gating transistors, energy is consumed during active-idle transitions due to discharging and charging of the capacitances at the virtual supply and internal circuit nodes by adder leakage currents as they converge to steady-state values. This "leakage convergence" time is measured to be 1μs at 75°C (Figure 8-5(a)) and increases to 10ms when an external 133nF decap is added to the virtual supply. While adding low-leakage decap to the virtual supply alleviates the performance impact of power gating transistors, the transition energy overhead becomes larger. As a result, the overall leakage power savings for 10μs idle time is only 30%, instead of 84% (Figure 8-5(b)).

However, if the leaky MOS decap which accounts for 10% of the adder area is moved from the real supply to the virtual supply, decap leakage is virtually eliminated during idle period since the virtual supply collapses to 0.4V. Then overall leakage power savings by power gating transistor improves from 80% to 90%, in spite of the additional transition energy overhead arising from extra capacitance at the virtual supply (Figure 8-6(a)). The idle leakage power also depends on the input vector and clock mode of the domino adder (Figure 8-6(b)). Therefore, effectiveness of these leakage control techniques can be improved by loading, during active to idle transition, an input vector or clock mode that provides the smallest leakage power.

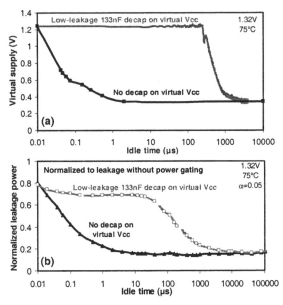

Figure 8-5. (a) Convergence of virtual V_{cc} as PMOS power gating (sleep) transistor is turned off. (b) Effect of virtual supply capacitance on time constant and leakage savings.

The minimum "idle" time required to achieve overall power saving is also dictated by the energy spent in switching the power gating transistor gates and body nodes. We compare total active power of these techniques at 4GHz clock frequency where higher supply voltage (1.32V) is used for the power gating transistor to meet the frequency target. Power measurements for different adder activity profiles, including switching energy overheads, show that the minimum "idle" time is ~100 clock cycles for both power gating transistors and body bias when the activity factor (α) is 0.05 (Figure 8-7(a)).

For 400 consecutive active cycles, the maximum activity factor, beyond which no power reduction is achieved, is 0.1 to 1 (Figure 8-7(b)). With wider execution pipelines and instructions spending larger amounts of time waiting for memory access, activity factors of execution units in high performance microprocessors are becoming smaller. Therefore, the minimum number of idle cycles or maximum activity factor limits for power gating transistor and body bias techniques to be effective (Figure 8-8(a)) are within the range of values encountered in microprocessors. Measurements for a typical activity profile demonstrate 10% overall active power reduction by power gating transistor and body bias (Figure 8-8(b)), in addition to that achieved by clock gating.

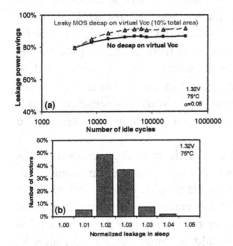

Figure 8-6. (a) Leakage savings as a function of decoupling capacitor placement. (b) Dependence of power gated leakage on adder input vector. Leakage measured after 100ns in sleep mode.

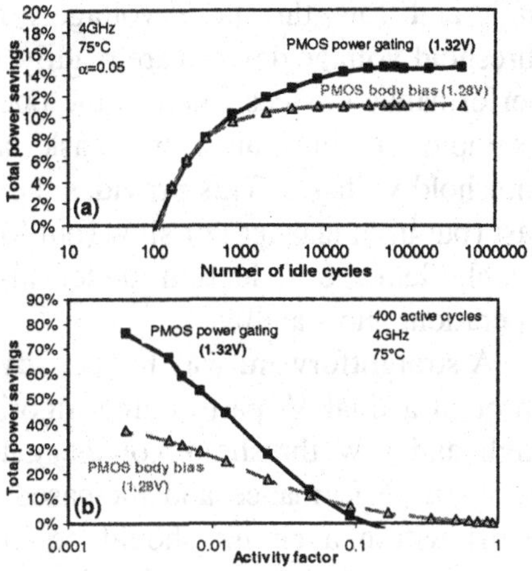

Figure 8-7. (a) Total power savings for power gating (sleep) transistor and body bias as compared to clock gating only for a fixed activity factor as a function of the number of consecutive idle cycles. (b) Total power savings as a function of activity factor for 400 consecutive active cycles.

(a)	Break-even number of idle cycles with $\alpha=0.05$	Break-even α with 400 active cycles
Adder with PMOS power gating transistor	115	0.13
Adder with PMOS body bias	121	> 0.5

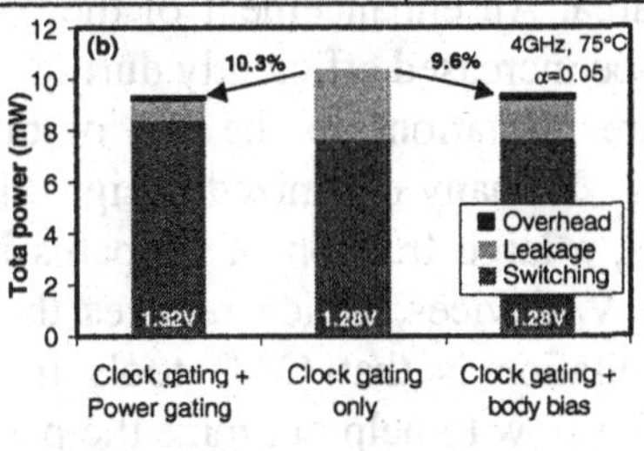

Figure 8-8. (a) Break-even point for total power in terms of number of idle cycles and activity factor. (b) Components of total active power for power gating (sleep) transistor and body bias, compared to clock gating only with 400 consecutive active cycles.

8.3 MULTI-PERFORMANCE DEVICES

Sub-threshold leakage currents are exponentially dependent on device V_t, and can be changed by several orders of magnitude by switching between high and low threshold voltages. In many modern processes, multiple threshold voltage devices are readily available to the circuit designer. Multi-performance devices, such as the ones achieved from a dual V_t process, for example, requires an extra mask layer to select between high and low threshold voltages. This provides the designer with transistors that are either fast (but high leakage) or slow (but low leakage). Multi-performance devices enable leakage reduction under all conditions, such as, testing, active operation, and standby.

A straightforward way to take advantage of these modern technologies is through a dual V_t partitioning algorithm. A circuit can be partitioned into high and low threshold voltage gates or transistors, which will tradeoff between performance and increased leakage currents. For instance, critical paths within a circuit should be implemented with low V_t to maximize performance, while non-critical paths should be implemented with high V_t devices to minimize leakage currents. By using fast leaky devices only when necessary, leakage currents can be significantly reduced in both the standby and active modes compared to an all low V_t implementation.

Dual V_t partitioning is a popular leakage reduction technique because the circuit operation remains the same as for a single V_t implementation, yet critical parts of a circuit can use scaled V_t devices to maintain performance at low supply voltages. [4, 5] discuss some examples that illustrate the effectiveness of this technique. It has also been shown in [6] that joint optimizations of dual-V_t allocation and transistor sizing reduce low-V_t usage by 36%-45% and leakage power by 20%, with minimal impact on total active power and die area. An enhancement of the optimum design allows processor frequency to be increased efficiently during manufacturing.

Practically, there are limitations to the use of dual V_t partitioning to reduce leakage currents. In many optimized designs there are many critical delay paths. Therefore, a large fraction of all paths in the circuit must be implemented with low V_t devices, which reduces the effectiveness of this technique. Another limitation is that CAD tools must be developed and integrated into the design flow to help optimize the partitioning process. It is not straightforward to identify which gates can be made high and low V_t without changing the delay profiles of the circuit. For example, one partitioning scheme that can be applied to random combinational logic is to first implement the circuit with all low V_t devices to ensure the highest possible performance, and then to selectively implant non critical gates to be high V_t. However, non critical gates which are converted to be high V_t

devices can become critical gates as illustrated in Figure 8-9. Additionally, due to process variation that is not well understood prior to manufacturing, assumption on critical and non-critical paths may end up being incorrect post manufacturing.

Significant research is still required to improve multiple V_t algorithms, especially in the presence of process variation in nanoscale CMOS. There exists a natural limit where dual V_t partitioning may not reduce standby leakage currents enough for ultra low power, high performance applications. As a result, other leakage reduction techniques are also important in nanoscale CMOS systems.

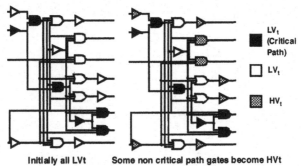

Figure 8-9. Dual V_t partitioning. Only some non critical path gates can be made high V_t.

REFERENCES

[1]   J. Tschanz, S. Narendra, Y. Ye, B. Bloechel, S. Borkar, and V. De, "Dynamic sleep transistor and body bias for active leakage power control of microprocessors," *IEEE Journal of Solid-State Circuits*, vol. 38, pp. 1838-1845, Nov. 2003.

[2]   N. A. Kurd et. al., "A multi-gigahertz clocking scheme for Intel® Pentium® 4 microprocessor." *IEEE Journal of Solid-State Circuits*, vol. 36, pp. 1647-1653, Nov. 2001.

[3]   S. Vangal. et. al, "5GHz 32b integer-execution core in 130nm dual-VT CMOS." *IEEE Journal of Solid-State Circuits*, vol. 37, pp. 1421-1432, Nov. 2002.

[4]   W. Lee, et al., "A 1V DSP for Wireless Communications," IEEE International Solid-State Circuits Conference, pp. 92-93, Feb. 1997.

[5]   T. Yamashita, et al, "A 450 Mhz 64b RISC Processor Using Multiple Threshold Voltage CMOS," *IEEE International Solid-State Circuits Conference*, pp. 414-415, Feb 2000.

[6]   J. Tschanz et. al., "Design Optimizations of a High Performance Microprocessor Using Combinations of Dual-V_t Allocation and Transistor Sizing," *Symp. of VLSI Circuits*, pp. 218-219, June 2002.

Chapter 9

IMPACT OF LEAKAGE POWER AND VARIATION ON TESTING

Ali Keshavarzi[§] and Kaushik Roy[¶]
§Intel Corpration, USA and ¶Purdue University, USA

9.1 INTRODUCTION

CMOS circuits by its construction consume very little power in standby. The concept of IDDQ testing, validating circuits by measuring and observing their quiescent supply current and its application to CMOS circuits was demonstrated by Mark W. Levi in his ITC'1981 paper titled "CMOS is most Testable." By definition IDDQ relies on the fact that CMOS circuits with increased leakage current are defective.

Technology scaling challenges the effectiveness of current-based test techniques such as I_{DDQ} testing because of elevated leakage and increased variation which have been discussed extensively throughout this book. Leakage and variation impact IC testing, burn-in and IC production yield while increase the manufacturing cost and packaging requirements severely. Furthermore, existing leakage reduction techniques e.g. applying reverse body bias (RBB) are not as effective in aggressively scaled technologies to address the current testing requirements. In this chapter, the concept of correlative and multi-parameter test techniques are investigated and studied. We are interested to fundamentally distinguish intrinsic (physics based) transistor (including gate leakage) leakage components contributing to chip leakage power from defect induced components.

In our effort toward this objective, we exploit intrinsic dependencies of transistor and circuit leakage on clock frequency, temperature, and reverse body bias (RBB) to discriminate fast ICs from defective ones. Transistor and circuit parameters are correlated to demonstrate a leakage-based test solution with improved sensitivity. We demonstrate the test techniques by experimentally measuring test ICs with available body terminals. Our data suggest adopting a sensitive multiple-parameter test solution. For high-performance IC applications, we propose an I_{DDQ} versus F_{MAX} (maximum operating frequency) test solution, in conjunction with temperature (or RBB) to improve the defect detection sensitivity. For cost sensitive applications, I_{DDQ} versus temperature test can be deployed. Experimental results in this chapter show that temperature (for example cooling from 110°C to room) improved sensitivity of I_{DDQ} versus F_{MAX} two-parameter test by more than an order of magnitude (13.8X). The sensitivity can also be tuned by proper selection of a temperature range to match a required defect per million (DPM) level.

Another challenge of elevated leakage is the ability to deal with power consumption under burn-in conditions. A brief overview of the problem and use of body bias to solve the problem is discussed at the end of this chapter.

In most tests performed exhausting all test vectors and conditions is not possible. It is assumed that a smaller sub-set will provide the necessary test coverage. This assumption may not be valid with increasing parameter variation. Additionally, most test methods assume ability to correlate one set of parameter to another set to arrive at conclusions – for example, it maybe assumed that if a wafer level test at 25°C and 1.2 V operates at 2 GHz, then it is expected that once packaged it will be able to operated at 2 GHz at 110°C and 1.35 V. Validating or arriving at such assumptions under the presence of parameter variation and leakage is a time intensive task. While this topic is not covered in-depth in this chapter, this is or soon expected to be an active area of research.

9.2 BACKGROUND

There is significant research that describes the impact of technology scaling on various aspects of VLSI testing [1-7]. The ability to perform leakage-based tests is threatened by elevated transistor leakage in scaled process technologies. This trend challenges conventional I_{DDQ} test methods. Recently, several methods were reported to sustain the effectiveness of I_{DDQ} testing for sub-0.25 μm technologies. These methods include reverse body bias (RBB), current signatures, delta I_{DDQ} testing, and transient current testing [2,3,9,16-20]. By applying a reverse body bias, a low leakage I_{DDQ} test mode is created [2,3]. Gattiker et al. suggested sorting of I_{DDQ} test vectors in ascending order where an abrupt discontinuity in the current level is an indication of a defect [16]. Maxwell et al. demonstrated the effectiveness of current signatures with silicon data [17]. Thibeualt [18] and Miller [19] proposed delta I_{DDQ} test technique to uncover defects.

Conceptually, delta I_{DDQ} technique is similar to that of current signatures where sudden elevation in current level is an indication of a defect. Some researchers suggested utilization of transient current test techniques [9,20]. The search for new test strategies is motivated primarily by the increasing adverse device leakage trends.

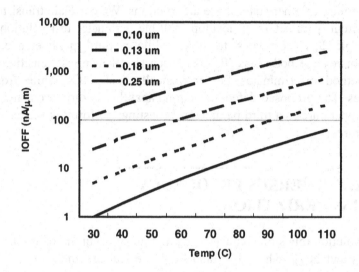

Figure 9-1. Projected transistor off-state current (I_{OFF}) [10].

We will illustrate one aspect of the problem of elevated intrinsic leakage by estimating the sub-threshold transistor leakage component and computing leakage of future chips [10]. We start with the 0.25 µm technology [11] and project sub-threshold leakage currents for 0.18 µm, 0.13 µm, and 0.1 µm technologies. Sub-threshold leakage is not the only component of a transistor leakage, but the sub-threshold leakage is generally the dominant component [3,4]. Even in scaled technologies that transistor gate leakage is significant, the transistor is designed such that sub-threshold leakage is dominant. A typical 0.25 µm transistor with V_T of 450 mV has an I_{OFF} of ~1nA/µm at 30 °C (room temperature). If sub-threshold slopes are 80 and 100 mV/decade at 30 °C and 100 °C respectively, and if V_T changes by 0.7 mV/°C and scales by 15% per technology generation, then I_{OFF} increases by 5X for each new technology generation [10].

Since I_{OFF} increases with temperature, it is important to consider leakage currents and leakage power as a function of temperature. Figure 9-1 shows I_{OFF} projected across four different technologies as a function of temperature [10]. I_{OFF} at room temperature increases from 1 nA/µm for a 0.25 µm technology to larger than 100 nA/µm for a 0.10 µm technology. At 110 °C, these values correspond to 100 nA/µm and greater than 1 µA/µm, respectively. We may use these projected I_{OFF} values to estimate the active leakage of a 15 mm² die integrated circuit. The total transistor width on the die is expected to increase by ~50% for each technology generation. Hence, the total leakage current increases by ~7.5X.

Our approach to testing intrinsically leaky integrated circuits is different, yet complementary to other state-of-the-art solutions. We correlate transistor device and circuit parameters to develop multiple parameter test solutions. As an example I_{DDQ} correlates to F_{MAX} while a third parameter e.g. temperature or reverse body bias (RBB) is used to enhance test sensitivity. This test method discriminates fast intrinsically leaky ICs while from defective ones. The proposed defect detection capability is further enhanced by adding temperature as a third parameter for testing (performing the test at two temperatures).

9.3 LEAKAGE VERSUS FREQUENCY CHARACTERIZATION

Our test solution relies on characterizing and quantifying the relationship between I_{DDQ} and F_{MAX}, where I_{DDQ} and F_{MAX} are leakage and chip speed, respectively. Keshavarzi et al., [3] correlated leakage current (I_{DDQ}) of a 32-bit microprocessor to its maximum clock frequency (F_{MAX}) in a 0.35 µm technology. This relation was reexamined for 0.18 µm technology test chip circuit. The test IC provided more flexibility and degrees of freedom such as availability of body terminals to explore the fundamental concepts prior to actual product specific test development. The roots of leakage to frequency correlation is based on device physics and for a circuit designed on a given process technology, one can establish and characterize such a relationship [3].

We used our high-performance test chip containing 20,000 transistors to study the relationships between I_{DDQ} and F_{MAX}. We studied dependencies of I_{DDQ} and F_{MAX} to Reverse Body Bias (RBB), transistor threshold volatge (V_T), transistor channel length (L), power supply voltage (V_{DD}), and temperature (T) for various technologies. Figure 9-2 is a photo of the test chip showing multiple circuit blocks. We used a ring oscillator (RO) and a delay chain circuit for our studies. The body (well and substrate) terminals of these devices on the ICs were externally available for RBB application. RBB is the potential between the body and source terminal that reverse biases the body to source *pn* junction. The body of the *p*MOS (*p*-channel) device is located in the *n*-well and the body of the *n*MOS (*n*-channel) device is linked to the *p*-well and to the *p*-substrate. The technology was a twin-well CMOS process with the *p*-well inside the *p*-substrate. All *n*-channel transistor bodies have the same body voltage when a bias is put on the substrate.

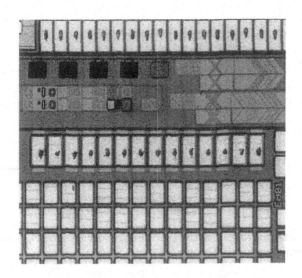

Figure 9-2. Die photograph of the test chip.

Figure 9-3 is a semi-log plot of log I_{DDQ} versus F_{MAX} without applying any RBB (zero body bias or RBB = 0 V) in our 0.18 μm technology. This semi-log curve plots the relationship between test chip ring oscillator (RO) normalized I_{DDQ} leakage and its normalized maximum operational frequency, F_{MAX}. Normalization is with respect to the lowest chip leakage and frequency (Figure 9-3). Data came from more than 100 ICs from two wafers where each data point in Figure 9-3 represents a die.

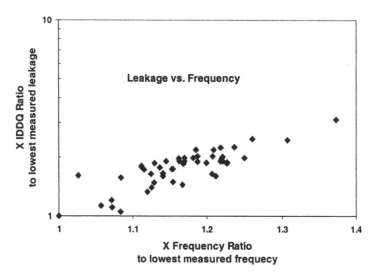

Figure 9-3. Normalized Ring Oscillator (RO) circuit frequency versus I_{DDQ} leakage at room temperature without any applied body bias.

The data indicates a linear relationship between log I_{DDQ} and RO F_{MAX}. This linear dependency is observed across the range of natural variation in transistor and circuit parameters in our ICs. No parameter such as V_T or L was intentionally skewed, hence, we observe the natural range of parameter variation in our experiment. For this collection of IC's, a 35% increase in CUT (circuit under test) F_{MAX} (1.35X change) results in an increase in I_{DDQ} by 4.1X at room temperature. Alternatively, the CUT that was faster by 35% also had 4.1X higher leakage. Slower (faster) circuits have lower (higher) leakage. This fundamental relationship between IC's maximum operating frequency (F_{MAX}) and its aggregate leakage (I_{DDQ}) is based on physics and is essential to developing the concept of two-parameter testing. The slower ICs have higher threshold voltage (and longer channel length) and hence leak exponentially less. Variation in transistor channel length and control in transistor critical dimension (CD) impact leakage exponentially while change delay linearly. This is consistent with transistor physics and short channel effects (SCE).

9.4 MULTIPLE-PARAMETER TESTING

Let us consider a universal multi-parameter test solution based on the intrinsic leakage to frequency correlation. Multiple-parameter testing correlates a parameter such as circuit leakage against another parameter such as circuit speed (operating frequency) or temperature and uses a third variable for example reverse body bias (RBB) or temperature to enhance the test sensitivity. Multiple-parameter testing is a low-cost alternative solution for discriminating fast intrinsically leaky ICs from defective ones. It is a method to extract a defect signal from a variable and noisy leakage to frequency dependence. To explore this further, since ICs with high levels of intrinsic leakage behave differently than defective ICs in context of I_{DDQ} vs. F_{MAX} test, one may use this information to develop a sensitive test method.

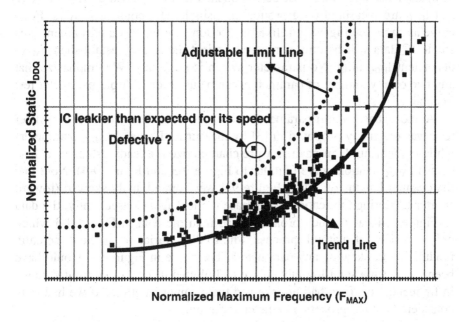

Figure 9-4. Example Microprocessor I_{DDQ} versus F_{MAX} with trend line and test limit [3].

During characterization and test development, leakage (I_{DDQ}) and maximum operating frequency (F_{MAX}) are measured and plotted against each other for many ICs. A trend line is established and superimposed on the measured data in Figure 9-4. A frequency dependent leakage limit line is also shown in Figure 9-4 as a dashed line. The leakage limit which is frequency dependent is determined by statistical analysis in order to provide the appropriate guard band to account for normal range of variability in data. The dependency trend line defines the shape of a limit line (Figure 9-4). Then, we may make a decision on a measured IC depending on where it lies on Figure 9-4 with respect to the proposed adjustable frequency (F_{MAX}) dependent leakage (I_{DDQ}) limit line. If for a given frequency, an IC has substantially higher leakage than forecast by the intrinsic dependency (similar to the circled IC in Figure 9-4), then the IC is classified as defective.

However, the IC identified in Figure 9-4 is a questionable IC because its leakage is not substantially higher. This IC is a candidate for further examination.

Thus, the two-parameter test limit should distinguish fast and slow ICs from those that are defective. The test improves signal to noise ratio for defect detection for high-performance ICs with high background (intrinsic) leakage levels. If the decision is doubtful, we may seek other variables to enhance the defect discrimination sensitivity of this test. The frequency dependent leakage limit must be established *a priori*. Then each IC's F_{MAX} and I_{DDQ} are measured once producing a single data point in I_{DDQ} vs. F_{MAX} curve as shown in Figure 9-4. In a production sort and test environment, this single IC data will be only compared against the already established F_{MAX}-dependent I_{DDQ} limit. No parameter sweep is necessary. We emphasize that any two or any number of parameters can be used for multiple-parameter test method.

A more basic multi-parameter microprocessor test, the two-parameter test, was originally proposed in [3] by measuring I_{DDQ} and F_{MAX} parameters while comparing them against a pre-characterized, already-established I_{DDQ} versus F_{MAX} relation curve (similar to Figure 9-3 and Figure 9-4). Note that the channel lengths of ICs in Figure 9-4 were intentionally skewed (toward shorter L) during fabrication to increase the leakage. Consequently, the data in Figure 9-4 has a much broader range than the data in Figure 9-3 which only consists of natural process (die-to-die) variation without actually modulating or skewing any parameters. The curve in Figure 9-4 would have been linear (in a semi-log plot) if we had plotted it for a more limited range in frequency, or if we had not skewed the channel length, or if we had only considered natural process parameter variation.

9.5 SENSITIVITY GAIN WITH RBB AND TEMPERATURE

Since the leakage (sub-threshold and gate leakage) to frequency correlation relation is inherent to the device physics, varying transistor, circuit, and environment parameters such as temperature and body bias cause predictable changes in this relationship/dependency. We used this concept to improve the sensitivity of the two-parameter test solution. Applying a reverse body bias (RBB) lowers IC leakage and reduces IC performance [3,4]. Lower temperature proves the transistor performance and circuit switching speed while reduces its leakage current. In this section, we show that applying temperature and/or reverse body bias (RBB) as a third parameter improves the signal to noise ratio of the proposed I_{DDQ} versus F_{MAX} test.

Figure 9-5 plots normalized RO (as the CUT) frequency change against its leakage as the reverse body bias voltage, V_{bs}, is varied from 0 V to 1.5 V in steps of 250 mV for two different temperatures. We used absolute values of the RBB as we applied the RBB to both *n*MOS and *p*MOS transistors in the RO. Data in Figure 9-5 is from a typical RO die. The lower curve is at room temperature (27.7 °C). As temperature rose to 110 °C, the leakage increased by about 30-40X; however, the frequency reduced by about 5-8%. Curves at both temperatures show the reduced leakage as RBB magnitude increases (going from right to the left along each curve). Moving from right to left along the T=27.7 °C curve, only a modest 2X reduction in leakage occurs for a 0.5 V RBB (second data point with a square around it in Figure 9-5). Furthermore, a 0.5 V RBB value appears to be an optimum value as the leakage reduction saturates beyond this point.

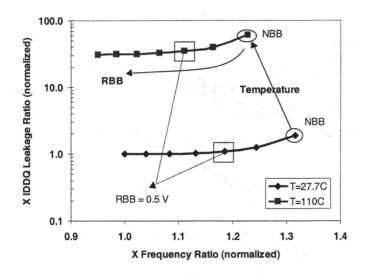

Figure 9-5. A typical RO leakage versus frequency at room temperature (27.7 °C) and at hot (110 °C) with varying reverse body bias from 0 to 1.5 V in steps of 250 mV.

Figure 9-5 shows that lowering the temperature from 110 °C to 27.7 °C changed the RO leakage much more than varying RBB from no bias to effectively 1.5 V, highlighting the sensitivity of I_{DDQ} vs. F_{MAX} relationship to temperature. Consequently, temperature enhances the two-parameter test sensitivity while RBB provides minimal leakage reduction resulting in a limited application for this 0.18μm technology [14]. It should be noted that RBB has minimal impact on gate leakage and for technologies where gate leakage is higher, RBB will be even less effective. However, sub-threshold leakage is very sensitive to temperature while gate leakage is not temperature dependent. Hence, sub-threshold leakage being the dominant leakage at higher temperature, RBB might be more effective at higher temperature.

Despite RBB's limited effectiveness in reducing leakage of scaled technologies, we studied its use as a third variable to improve the sensitivity of two-parameter test for defect discrimination. RBB alters the fundamental I_{DDQ} versus F_{MAX} relationship and statistics. Figure 9-6 shows the shift in RO leakage and frequency by applying 0.5 V of RBB to ICs of Figure 9-3. The arrows shown in Figure 9-6 pictorially represent the direction of the shift in speed and leakage as a result of applying reverse body bias. On average, for all ICs tested, a 0.5 V RBB resulted in leakage reduction of 1.8X while speed reduced by 10% (1.1X).

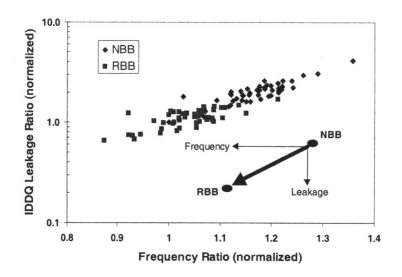

Figure 9-6. IC leakage versus frequency with and without reverse body bias of 0.5 V at room temperature (27.7 °C).

Next we re-analyzed and evaluated the data in Figure 9-6 by only considering leakage reduction from applying a reverse body bias voltage (Figure 9-7). If we ignore the frequency shift (slow-down) associated with RBB, we can plot the IC's leakage before and after applying reverse body bias as a function of the IC's original frequency with RBB = 0 V. Thus, the frequency identifies each die or IC allowing for tracking of leakage with applying reverse body bias. Figure 9-7 shows data plotted in this manner (same data set as Figure 9-6). The vertically paired points are the two bias conditions for the same IC. The arrow in Figure 9-7 shows the direction of leakage reduction with RBB. The x-axis represents the original IC F_{MAX} i.e. with no body bias applied (NBB).

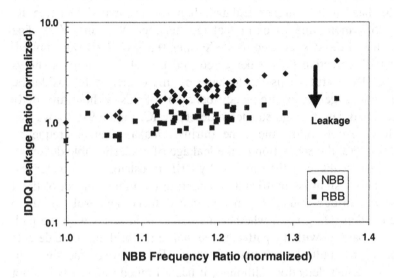

Figure 9-7. IC leakage with and without reverse body bias of 0.5 V versus IC's original frequency (i.e. no body bias) at room temperature (27.7 °C).

Figure 9-7 more clearly shows that reverse body bias reduces IC leakage creating a new leakage versus frequency correlation curve separated from the original dependency with a different slope. Physically this shift is caused by the increase in V_T due to applied reverse body bias. This suggests that leakage of defect-free intrinsically leaky ICs can be changed in a controlled manner by body bias. However, if an IC is defective and the defect creates a parallel leakage path between V_{DD} and V_{SS}, the shift in current will be different from the intrinsic physics-based behavior. We basically expect the defect to be independent of RBB. Thus, the defective ICs I_{DDQ} does not reduce as much as an intrinsically leaking IC. This property can improve the sensitivity of the proposed two-parameter testing.

Emulating a small defect, a large 4 MΩ resistor is placed between V_{DD} and V_{SS} of two of our ICs. For V_{DD} = 1 V, this resistor gives 25 nA of leakage path in parallel to our IC under test (CUT). The data in Figure 9-7 is re-plotted in Figure 9-8 with two ICs having 4 MΩ rail bridge defects. The 4 MΩ resistance value is selected due to the relatively low leakage in our 20,000 transistor test chip and it provided a defect with very low impact on signal disturbance. The objective is to show how, even subtle, defects can be screened. Gross defects are otherwise detectable by the adjustable limit concept (a frequency-dependent leakage limit) in the original two-parameter testing. One does not need to use a third variable to improve the test sensitivity for catching gross defects.

The emulated defective IC in the middle of Figure 9-8 shows that the defective IC has slightly higher leakage than the original defect-free IC (1.25X or 25% more due to extra leakage between V_{DD} and V_{SS}). The median leakage of the ICs reduced by 1.8X when 0.5 V of RBB was applied. The "emulated" defective IC's leakage reduced by 1.45X. The data shows that 0.5 V of RBB still keeps the leakage of this defective IC inside the population of leakage versus frequency behavior of ICs without any body bias. Hence, RBB can distinguish such a defective IC. Figure 9-8 shows that defective ICs do not follow the same intrinsic leakage versus frequency slope. Furthermore, the separation of the leakage of a questionable defective IC from intrinsic leakage of the same IC by RBB translates into a better test sensitivity. This insight is significant considering the minor impact of the 4 MΩ rail to rail bridge defect. Often resistive defects can result in subtle delay failures, depending upon whether the defect is in series with the path e.g. a via or in parallel with the path e.g. a signal to rail bridge. These defects can be isolated by multiple-parameter testing. RBB improves the signal to noise ratio for defect detection; although, it has a limited range. RBB is not able to fully separate out the leakage versus frequency curves.

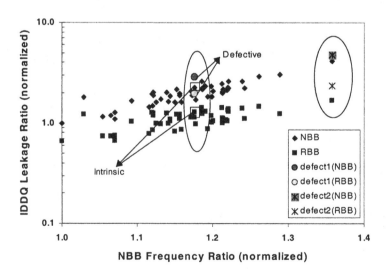

Figure 9-8. Defective and defect-free IC leakage with and without reverse body bias versus IC's original measured frequency (with no applied body bias).

Additionally, we emphasize that RBB leakage reduction and its application in improving test sensitivity is diminishing as the technology scales [12-13]. One limitation of RBB comes from GIDL (Gate Induced Drain Leakage) and reverse biased junction leakage [12-13] -- primarily due to band-to-band tunneling. Another limitation is degradation of the body effect due to short channel effects (SCE) which reduces the body potential modulation of V_T substantially in scaled technologies. Results from a 0.18 μm technology suggest a different conclusion than reported for an older 0.35 μm process technology [3]. Reverse body bias lowered leakage and hence I_{DDQ} by three orders of magnitude in a 0.35 μm technology [3]. However, recent data showed that the effectiveness of RBB to lower I_{OFF} decreases as technology scales. Keshavarzi et al., showed that RBB could only reduce leakage by less than a single order of magnitude for scaled 0.18 μm technology [12-13]. Technology parameters, particularly junction doping concentrations and dimensions, play a major role in the effectiveness of RBB [12-13].

Looking at temperature as another parameter to improve test sensitivity, Figure 9-9 plots leakage versus frequency of the same ICs as a function of two temperatures. The arrow in Figure 9-9 shows the direction of leakage and speed change as ICs cool down. Figure 9-10 plots the same data in Figure 9-9, but ignores the frequency shift of changing the temperature. In other words, we plotted the change in leakage as a function of IC's original frequency at room temperature. It is apparent that temperature is more effective in modulating the leakage for the 0.18 μm technology than RBB. As technology scales, temperature may become more effective than RBB. For our technology, we could only approach ~2X leakage reduction on average for RBB (at the optimum body bias point of 0.5 V) as opposed to approximately ~32X by temperature (by cooling from hot to room temperature). This compares RBB and temperature in order to apply these leakage modulation techniques for test sensitivity enhancement. Note that we did not purely rely on the magnitude of I_{DDQ} leakage reduction in these leakage modulation methods, but we used them for sensitivity purposes in two-parameter test.

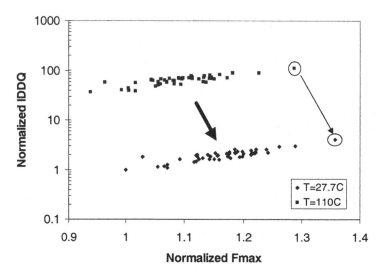

Figure 9-9. IC leakage versus frequency at room temperature (27.7 °C) and at hot (110 °C) without any reverse body bias.

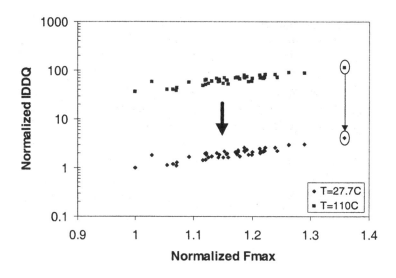

Figure 9-10. IC leakage versus IC's original frequency (i.e. with no body bias at room temperature) for two different temperatures (27.7 °C and 110 °C).

We measured temperature dependence of our ICs for a wide range of temperature, from –50 °C to 110 °C in six steps. The intrinsic leakage versus frequency data shown in Figure 9-11 suggest that we can change the leakage of our ICs by more than three orders of magnitude in this temperature range. We also see that ICs are slowing down as the temperature increases by a shift of the data to the left in Figure 9-11. This data shows that temperature is effective in improving the sensitivity of the proposed two-parameter test.

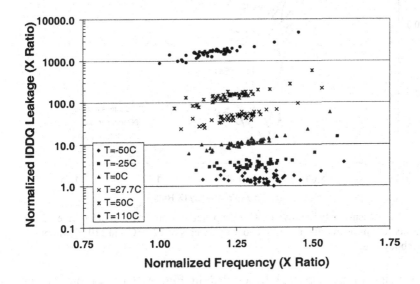

Figure 9-11. IC leakage versus frequency for different temperatures without any reverse body bias.

We emulated a defective IC by adding a bridge (1 MΩ resistor) between V_{DD} and V_{SS} which is similar to the technique used for RBB analysis. Figure 9-12 shows a good separation between intrinsic population of ICs at two temperatures, studied for defect sensitivity improvement. No intrinsic data overlap due to natural frequency versus leakage variation occurred in the frequency versus leakage data at two temperatures of 27.7 °C and 110 °C. Comparing Figure 9-12 to Figure 9-8, temperature is very effective in distinguishing "defective" and "good" ICs.

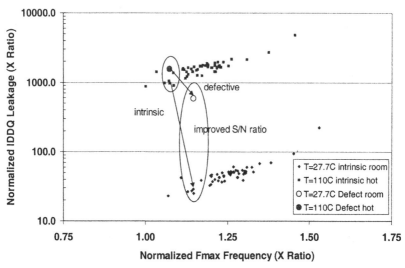

Figure 9-12. Defect sensitivity improvement by temperature. Defective and defect-free IC leakage versus IC's measured frequency at two temperatures of 27.7 °C and 110 °C without any applied body bias.

The leakage of the original IC circled in Figure 9-12 for our defect sensitivity study reduced intrinsically by 36X when temperature dropped from "hot" to "room" temperature. When we added the defect to this IC, the leakage of this "emulated" defective IC shown in Figure 9-12 increased by 1.6X due to the extra leakage path between V_{DD} and V_{SS}. This data point still belonged to the hot leakage versus frequency population plot making it a challenge to detect this defect purely by the adjustable limit concept. In other words, two-parameter testing with the proposed adjustable frequency-dependent leakage limit lacked the necessary sensitivity to detect and isolate this defective IC. When the temperature was lowered, however, the leakage of this defective IC reduced by only a factor of 2.6X, keeping this defective IC outside of the main population of frequency versus leakage behavior at lower (room) temperature. We quantified the gain in the test sensitivity by taking the ratio of intrinsic leakage reduction (36X) to the amount of leakage reduction for the defective IC (2.6X). The signal to noise ratio improved by more than an order of magnitude (36X/2.6X=13.8).

Figure 9-12 shows that the defective IC's leakage at room temperature is still located at about the leakage versus frequency dependence distribution at hot temperature. Consequently, a simple adjustable limit concept applied to this data point at room temperature can now detect this defect. Lowering temperature by 80°C (from hot to room temperature) provided an order of magnitude separation (approximately 40X versus 4X) between intrinsic

(40X) and defect-induced (4X) amount of leakage reduction in the context of two-parameter testing. This is primarily the reason why temperature improves sensitivity and signal to noise ratio for the two-parameter testing. Temperature in conjunction with the adjustable limit may detect more subtle defective ICs.

9.6 LEAKAGE VERSUS TEMPERATURE TWO-PARAMETER TEST SOLUTION

I_{DDQ} may be combined with temperature when F_{MAX} is not measured or is not relevant. An intrinsic two-parameter relationship between leakage and temperature can screen for defective ICs while considering the natural range of parameter variations. One may make two leakage (I_{DDQ}) measurements at two different temperatures. These are then mapped against a pre-characterized leakage versus temperature behavior. This is done for an IC product fabricated on a given process technology incorporating variation in temperature, frequency, and leakage. Figure 9-13 shows non-defective IC leakage as a function of temperature for a range of different temperatures from -50 °C to 110 °C. The spread in leakage data at each given temperature represents the natural spread resulting from parameter variations. This variation is similar to the natural range of parameter variations observed in leakage versus frequency behavior (Figure 9-3). This variation needs to be fully characterized in production worthy test solutions to select a proper temperature range.

Figure 9-13 also complements our earlier conclusion by showing that temperature as a good modulator of intrinsic leakage for our technology, over a range of our test chip frequencies and a range of parameter variations. The key in selecting the temperatures for testing is to make sure that the two temperatures are separated enough to ensure that the leakage variability resulting from parameter variations (spread in the data at a given temperature) do not to overlap. For example, lowering temperature from 110 °C to 27 °C reduces the mean intrinsic leakage by ~40X. And the natural spread in data does not overlap each other, providing a means for defect detection.

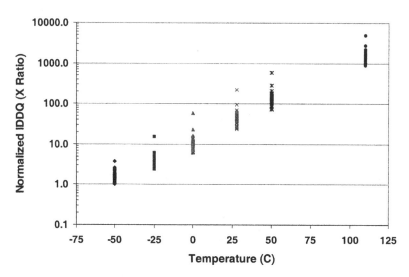

Figure 9-13. IC leakage versus temperature for six different temperatures without any body bias.

We demonstrated the defect detection capability for I_{DDQ} versus temperature test using a defective IC (same defective IC used in Figure 9-12) to show how this test improves the signal to noise ratio. Figure 9-14 shows that one can separate out a defective IC in a population of normal ICs at hot temperature by lowering the temperature to room. The defective IC's leakage at room temperature is outside the natural range of leakage parameter variation at the same temperature. The defective IC's leakage reduced by 2.6X while it should have been lowered intrinsically by 36X, which is an order of magnitude larger. The lower temperature enables better separation, resulting in a better signal to noise ratio. A wider temperature separation also allows for tolerating a wider range in the parameter variation. Furthermore, the sensitivity may be tuned to a desired level with a selection of a proper temperature range.

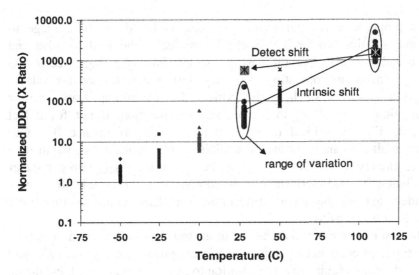

Figure 9-14. IC defect detection for leakage versus temperature two-parameter test at two different temperatures (27.7 °C and 110 °C) without any body bias.

9.7 DISCUSSIONS AND TEST APPLICATIONS

Leakage reduction techniques (e.g. lowering temperature, lowering V_{DD}, applying RBB, chip segmentations, etc.) by themselves may not be sufficient for IC leakage testing in the future scaled technologies. To pursue average leakage reduction for all ICs, one needs to aggressively combine several of these techniques to maintain a reasonable efficiency. Furthermore, they add to the cost of an already expensive test process. For example, very low temperature adds to the test cost and RBB adds to the design cost. Although we can use any leakage reduction we can get, we may have to look for other solutions for the problem of testing of ICs with elevated intrinsic leakage. The test proposed here views an IC's leakage (no matter how high) in the context of its maximum frequency (correlative). Consequently, it does not require an IC's average intrinsic leakage to be low making it a scalable test solution. It can also manage variability in leakage. Multiple-parameter testing is a low cost alternative for saving fast intrinsically leaky ICs while discriminating against defective ones. This test does not contradict any current test practices and measurement techniques. It actually complements all currently available techniques [2,3,9,16-20]. We defined two different test applications, one for the high-performance IC market and one for low cost applications.

9.7.1 High-Performance Products Test Applications

We proposed a two-parameter test solution based on the leakage to frequency relationship for high speed IC products fabricated on advanced process technologies. I_{DDQ} (at least in a simple form of a single vector I_{SB}) and F_{MAX} (maximum operating frequency) tests are performed for today's high-performance ICs. I_{DDQ} is critical in measuring static power consumption of an IC to the data sheet specification, if not for defect screening. F_{MAX} is critical to speed bin the high-performance ICs. Our primary multiple parameter test methodology utilizes these two parameters that are already measured. This forms the I_{DDQ} versus F_{MAX} two-parameter test solution for high-performance microprocessors. Adjustable frequency-dependent leakage limit and temperature (or RBB) enhances this test's defect detection sensitivity.

This test can be inserted at the end of the test flow (not consuming any IC tester time) prior to inking defective ICs or throwing away the packaged units. It does not require any modification to currently established test flows. Furthermore, it does not take additional tester time because it only requires off-line data processing. We used an IC's measured I_{DDQ} and F_{MAX} parameters and compared them against a pre-characterized, already-established I_{DDQ} versus F_{MAX} relation curve, similar to Figure 9-3 and Figure 9-4. This relationship is fundamental in defining a frequency-dependent leakage limit. If for a given frequency, an IC has substantially higher leakage than expected by the intrinsic dependency (similar to the IC shown in Figure 9-4), then the IC is classified as defective. If the decision is doubtful, we may apply reverse body bias (RBB) to the questionable IC only if a body terminal is available [15] or lower its temperature to measure its leakage. Our study shows temperature to be more effective for improving the test sensitivity.

There is a cost associated with improving sensitivity for low DPM applications since we introduced a third variable to the test. However, only leakage is measured. There is no need to re-measure the IC's F_{MAX} at different temperatures. This saves cost especially since the F_{MAX} test is expensive. F_{MAX} consumes test time because it requires applying appropriate test vectors while we step and shrink the clock period to determine at what frequency the IC stops working. Also, there is no need to sweep any parameters (e.g. F_{MAX}). We compare against *a priori* known speed-adjusted leakage limit characterized at two temperatures. If the leakage was reduced in line with a pre-determined (pre-characterized) intrinsically expected value, then the IC is not defective. If not, the IC will be classified as potentially defective. The separation and test limit at lower temperature screens the defective ICs. If the temperature range we use for sensitivity improvement is wide enough, then we can detect more subtle lower current conducting defects such as delay defects. Finally, the test is flexible and can be tuned to various applications requiring different degrees of quality, sensitivity, and DPM levels.

9.7.2 Test Applications for Low Cost Products

Some products do not measure F_{MAX} or are designed for a fixed frequency e.g., low cost products, fixed-speed ICs, ASICs, and constant throughput applications (DSPs). Here, one may use I_{DDQ} versus temperature to establish a correlative two-parameter test solution. Another approach is to add a ring oscillator (RO) circuit or any other delay chain circuits to the silicon white space for frequency measurement. Then we can measure this special circuit's frequency and plot it against actual IC product leakage for test purposes. Therefore, even this category of products can benefit from the advantages of I_{DDQ} versus F_{MAX} test.

9.7.3 Burn-in Testing

Burn-in testing is a method of reliability assurance testing that pre-tests a CMOS chip at elevated voltage and temperature conditions. Leakage power under these conditions can become prohibitive in CMOS chips that use leaky transistors. In high-performance systems, burn-in is often performed at very low clock frequency to conserve power consumption. CMOS chips that pass specific burn-in test condition would be expected to have a highly reliable functionality for a pre-specified duration under normal operating conditions. Elevated supply voltage and temperature under burn-in conditions exponentially increase the leakage power under such conditions compared to normal operation. It is expected that burn-in leakage power consumption may approach or even exceed the total power consumed under normal operating conditions.

Reverse body bias under burn-in testing can be used to reduce sub-threshold leakage power. It was shown discussed in Chapter 5 (Figure 5-13) that there exists an optimal reverse bias below which the power consumption increases due to increase in tunneling leakage. Also, this optimal reverse bias value is becoming smaller with scaling [12]. The threshold voltage modulation range to control leakage can be improved if the normal operation of the transistor requires forward bias value and under burn-in conditions the forward bias is withdrawn. This increases the leakage saving potential to more then triple compared to using the reverse bias, since forward bias has better control of the threshold voltage. Figure 5-20 in Chapter 5 showed that as much as 30X leakage reduction is possible at elevated temperature if forward body bias is used for normal operation and it is withdrawn under elevated temperature conditions. More leakage savings can be achieved if reverse bias is applied instead of just withdrawing the forward bias.

9.8 CONCLUSION

This chapter examined the impact of technology scaling on the testability of deep submicron CMOS ICs. Elevated leakage currents are an essential element of aggressively scaled devices and technologies. These elevated intrinsic leakage distributions challenge the effectiveness of I_{DDQ} based test techniques. We characterized transistor and circuit leakage by measuring their sensitivities across frequency, temperature, and body bias. The results show that fundamental device behavior can be used to improve the test sensitivity. Device leakage and its switching speed are functions of the threshold voltage (V_T) and transistor channel length (L or effective electrical length, L_{eff}). Therefore, a strong correlation and dependency can be established between the I_{DDQ} and F_{MAX} (maximum operational speed) of a collection of ICs.

Our characterization produced several test solutions for ICs in aggressively scaled technologies. Multiple parameter testing solutions that combine parameters such as I_{DDQ}, F_{MAX}, temperature, reverse body bias and lowering power supply voltage can prolong the usefulness of I_{DDQ} or other effected tests. High-performance microprocessors may use two-parameter I_{DDQ} combined with F_{MAX} test. This test method is a low cost alternative for saving fast intrinsically leaky ICs while discriminating defective ones. In other IC applications where F_{MAX} is not measured, one may use the two-parameter I_{DDQ} versus temperature test solution. We used RBB and temperature parameters as test variables to enhance the test sensitivity and have determined that temperature is the most effective parameter. We showed that temperature improved the test signal to noise ratio by an order of magnitude. The sensitivity can be tuned with a proper temperature range for meeting more stringent quality requirements.

Finally, the use of body bias to address burn-in leakage testing was highlighted.

REFERENCES

[1] G. Singer, "The Future of Test and DFT," *IEEE Design & Test of Computers*, July-September 1997, pp. 11-14.
[2] M. Sachdev, "Deep Submicron I_{DDQ} Testing: Issues and Solutions," *European Des. & Test Conf.*, pp. 271-278, March 1997.

[3] A. Keshavarzi, K. Roy and C.F. Hawkins, "Intrinsic Leakage in Low Power Deep Submicron CMOS ICs," *Int. Test Conf., pp.* 146-155, 1997.

[4] A. Ferre and J. Figueras, "I_{DDQ} Characterization in Submicron CMOS," *Int. Test Conf.,* pp. 136-145, Nov. 1997.

[5] W. Needham, C. Prunty, E.H. Yeoh, "High Volume Microprocessor Test Escapes, an Analysis of Defects our Tests are Missing," *Int. Test Conf.,* pp. 25-34, 1998.

[6] P. Nigh et al., "So What is an Optimal Test Mix? A Discussion of the Sematech Methods Experiment," *Int. Test Conf.,* pp. 1037-1038, 1997.

[7] T.W. Williams, R.H. Dennard, R. Kapur, M.R. Mercer, W. Maly, "I_{DDQ} Test: Sensitivity Analysis of Scaling," *Int. Test Conf.,* pp. 786-792, Oct. 1996.

[8] Semiconductor Industry Association Roadmap for Semiconductors, 1997.

[9] M. Sachdev, V. Zieren and P. Janssen, "Defect Detection with Transient Current Testing and its Potential for Deep Sub-micron ICs," *Int. Test Conf.,* pp. 204-213, 1998.

[10] V. De and S. Borkar, "Technology and Design Challenges for Low Power and High-performance," *1999 ISLPED,* pp. 163-168, August 1999.

[11] M. Bohr, et al., "A High-performance 0.25 um Logic Technology Optimized for 1.8V Operation," *IEDM Tech. Dig.,* p. 847, 1999.

[12] A. Keshavarzi, S. Narendra, S. Borkar, C. Hawkins, K. Roy, and V. De, "Technology Scaling Behavior of Optimum Reverse Body Bias for Standby Leakage Power Reduction in CMOS IC's," *1999 Int. Symp. On Low Power Electronics and Design,* p. 252, Aug. 1999.

[13] A. Keshavarzi, C. Hawkins, K. Roy, and V. De, "Effectiveness of Reverse Body Bias for Low Power CMOS Circuits," *8th NASA Symposium on VLSI Design,* pp. 2.3.1-2.3.9, Oct. 1999.

[14] S. Yang, et al., "A High-performance 180 nm Generation Logic Technology," *IEDM Tech. Dig.,* p. 197, 1998.

[15] H. Mizuno, K. Ishibashi, T. Shimura, T. Hattiri, S. Narita, K. Shiozawa, S. Ikeda, and K. Uchiyama, "An 18-uA Standby Current 1.8 V 200 MHz Microprocessor with Self Substrate Biased Data Retention Mode," *IEEE Trans. Solid State Ckt.,* pp. 1492-1499, Nov. 1999.

[16] A. Gattiker and W. Maly, "Current Signatures: Applications," *Int. Test Conf.,* pp. 156-165, 1997.

[17] P. Maxwell et al., "Current Ratios: A Self Scaling Technique for Production IDDQ Testing," *Int. Test Conf.,* pp. 738-746, 1999.

[18] C. Thibeault, "An Histogram Based Procedure for Current Testing of Active Defects," *Int. Test Conf.,* pp. 714-713, 1999.

[19] A. C. Miller, "IDDQ Testing in Deep Submicron Integrated Circuits," *Int. Test Conf.,* pp. 724 - 729, 1999.

[20] B. Kruseman, P. Janssen and V. Zieren, "Transient Current Testing of 0.25 um CMOS Devices," *Int. Test Conf.,* pp. 47-56, 1999.

Chapter 10

CASE STUDY: LEAKAGE REDUCTION IN HITACHI/RENESAS MICROPROCESSORS

Masayuki Miyazaki, Hiroyuki Mizuno, and Takayuki Kawahara
Central Research Laboratory, Hitachi Ltd.

10.1 LEAKAGE REDUCTION USING BODY BIAS IN A RISC MICROPROCESSOR

Several techniques for reducing leakage power are implemented in Hitachi Ltd. and Renesas Technology Corp.'s RISC microprocessors. Low-power processors are particularly important in mobile equipment because battery lifetime is critical. Leakage-power reduction is effective in extending this lifetime. Generally, a RISC microprocessor has two operating modes — an active and a standby mode. The processor executes operations in the active mode, while operations are halted in standby mode. The processor is required to keep data in standby mode, which may often be for relatively long periods in mobile applications. Leakage reduction in standby mode is therefore essential.

Body bias control is one of the most useful techniques for leakage reduction. The control technique described below was applied to SuperH-4 (SH-4) microprocessors. Two different SH-4 processors were used as test chips. One was a high-speed SH-4 processor with 200-MHz/1.0-W performance, and the other was a low-power SH-4 processor with 167-MHz/400-mW performance. A supply-voltage reduction technique called data-retention control was used in addition to body bias control.

10.1.1 Basic Concept of Using Body Bias to Reduce Leakage

The supply voltage (V_{dd}) and body bias (V_{bb}) control the DC characteristics of MOS transistors, i.e., the threshold voltage (V_{th}) of the transistors. The relation between the V_{th} and V_{bb} is described as

$$V_{th}(V_{bb}) - V_{th}(0) = \gamma\left(\sqrt{V_{bb} + 2\Phi_f} - \sqrt{2\Phi_f}\right) \tag{1}$$

where γ is the body bias effect coefficient and Φ_f is the Fermi potential. The V_{th} is roughly proportional to the square root of the V_{bb}.

The V_{bb} can be used to vary the DC-current features of MOS transistors, as shown in Figure 10-1. The sub-threshold leakage current is represented by the line where V_{bb} equals 0 V. When a -1.5-V reverse body bias is applied, the V_{th} increases and the leakage current decreases. However, an excess reverse bias of -2.3 V increases the leakage current because of a gate-induced drain-leakage (GIDL) effect [1]. Therefore, an excess reverse bias cannot be used to reduce leakage. In contrast, reducing the V_{dd} increases the V_{th} and decreases leakage because of a drain-induced barrier-lowering (DIBL) effect. There is a large reduction in leakage when 1.0-V V_{dd} and a -2.3-V reverse body bias are used simultaneously. We call this method data-retention control.

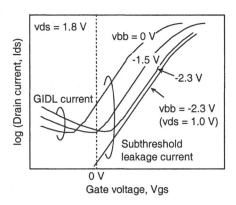

Figure 10-1. I_{ds}-V_{gs} characteristics of n-MOS transistor

10.1.2 Switched Impedance

Switch cells are designed to change the body bias in the active mode and the standby mode for body bias control. If the switch that connects the power supply and body bias produces high impedance, the performance of the processor is degraded by operating noise. The switch cells must therefore be carefully designed. The switched-impedance scheme varies the impedance between the power supply and body bias depending on the operating mode, as shown in Figure 10-2 [2]. The body bias control circuit (VBC macro) handles the V_{bb} of the 1.8-V logic area. V_{bp} and V_{bn} refer to the body bias for p-MOS and n-MOS transistors, respectively. VBCP and VBCN circuits supply reverse V_{bb} through high-impedance drivers in standby mode. A VBC macro connects a power source and body bias using the switch cells in the active mode. Up to 10,000 pieces of switch cells are distributed over the processor chip. The signals cbp and cbn control the switch cells. This switch structure lowers the impedance between the power source and body bias in the active mode. The VBCG is made from a charge-pump circuit and provides a negative voltage for V_{bn}. $V_{sub} = V_{bn} = V_{dd} - V_{well} = -1.5V$. The VBCR circuit is placed on the opposite side of the chip from the VBC macro. It detects transitions in body bias over the whole chip and the standby controller manages the operation of the 1.8-V logic area.

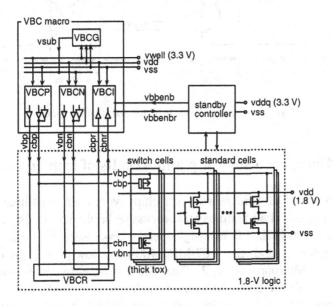

Figure 10-2. Switched impedance system

Signal transition between the active and standby modes is shown in Figure 10-3. 1.8-V V_{bp} and 0.0-V V_{bn}, that is, a 0-V body bias (V_{bb}), are applied in the active mode. Conversely, 3.3-V V_{bb} (=V_{well}) and −1.5-V V_{bn} (=V_{sub}), that is, a -1.5-V reverse V_{bb}, are applied in the standby mode. The standby controller generates a signal vbbenb and the vbbenb makes the decision to change the V_{bb}. A vbbenbr signal means that the mode transfer is completed for the whole chip. The transfer from the active to the standby mode takes 50 µs. This is a relatively long period because only the charge pump in the VBCG and buffers in the VBCP/VBCN drive the body bias lines. However, the long transfer period does not affect chip performance because the operation of the chip is halted after a transition to standby mode. When the mode returns to active, the switch cells enable a quick bias-voltage recovery within 370 ns.

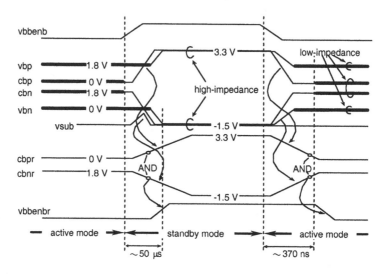

Figure 10-3. Signal levels in active and standby mode

10.1.3 Microprocessor Chip and Results of Implementation

Two different SH-4 microprocessor chips were used in estimating body bias control [2, 3]. Their features and process technologies are listed in Table 10-1. The 200-MHz/1.0-W high-speed processor is described as Microprocessor 1 and the 167-MHz/400-mW low-power processor is described as Microprocessor 2. The SH-4 has a two-way superscalar architecture with an 8-kB instruction cache and a 16-kB data cache. It contains direct memory access controllers (DMAC), timer units, serial communication interfaces (SCI), a real-time clock (RTC), bus controllers, and SDRAM interfaces. A 1.8-V V_{dd} is used for internal circuits and a 3.3-V Vwell is used for the other peripheral circuits and the body bias controller. The microprocessors include thin oxide thickness (T_{ox}) transistors and thick tox transistors. A 1.8-V V_{dd} is supplied to thin T_{ox} devices and a 3.3-V Vwell is supplied to thick ones. In a thin T_{ox} device, Microprocessor 1 has 0.2-μm gate length (L_g) transistors and Microprocessor 2 has 0.18-μm L_g transistors. A triple-well structure is used for body bias control. Here, we assume that V_{th}

is the gate voltage (Vgs) at a drain current (Ids) of 10-nA with a 15-μm MOS width (W). Micrographs of the chips are shown in Figure 10-4.

The effects of leakage reduction are shown in Figure 10-5. Microprocessor 1 (high-speed processor) reduced the leakage from 1.3 mA to 47 μA as a result of V_{bb} control. Data-retention control further reduced the leakage to 18 μA. With these controls, the V_{dd} was reduced to 1.0 V while preserving memory information, and the V_{bb} became −2.3-V (=V_{dd} – Vwell) reverse bias. The operating current of the VBC macros was 10 μA, which is included in the above standby current. The switched-impedance scheme was altered in Microprocessor 2 (low-power processor) to improve its performance. Figure 10-3 shows the signal levels of cbp and cbn reaching full amplitude of −1.5 V to 3.3 V. This design change enabled the scheme to operate at a lower V_{dd}. This processor reduced the leakage from 170 to 17 μA as a result of V_{bb} control.

Table 10-1. Microprocessor Features and Process Parameters

	Microprocessor 1 (high-speed version)	Microprocessor 2 (low-power version)
Architecture	2-way superscalar	
Cache	8-kB I-cache, 16-kB D-cache	
Power supply	1.8 V (V_{dd}), 3.3 V (V_{well})	
Transistor count	3.3 M	4.3 M
Clock frequency	200 MHz	167 MHz
Power consumption	1.0 W	400 mW
Standby leakage	1.3 mA (w/o body bias)	170 μA (w/o body bias)
	47 μA (with body bias)	17 μA (with body bias)
	18 μA (data retention)	
Area	6.8 x 6.8 mm (processor)	7.0 x 7.8 mm (processor)
	210 x 645 μm (V_{bb} cont.)	370 x 400 μm (V_{bb} cont.)
L_g	0.2 μm / 0.35 μm	0.18 μm / 0.35 μm
T_{ox}	4.5 nm / 8.0 nm	
V_{th}	0.15 V / 0.45 V	
Metal	5 layers	
Process technology	p-substrate, triple-well CMOS	

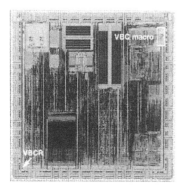 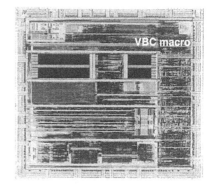

Figure 10-4. Chip micrographs: high-speed processor (left), low-power processor (right).

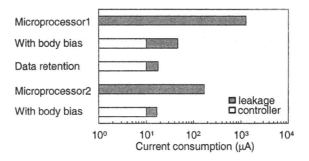

Figure 10-5. Measured current consumption

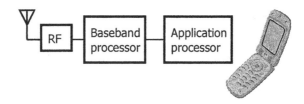

Figure 10-6. Application processor in 3G cellular phones

10.2 LEAKAGE REDUCTION IN APPLICATION PROCESSOR IN 3G CELLULAR PHONE

In this section we describe a method of reducing leakage power in an application processor in 3G cellular phones. 3G phones are used not only for voice communication, email, and web browsing, but also for more advanced functions such as videophone and 3D Java games. The phones have an application processor embedded them in addition to a base-band processor to achieve multimedia performance on demand without compromising standby or talk-time capacity (see Figure 10-6).

The challenge for chip designers is to maintain a long enough battery life to support these applications. A common solution is to provide several low-power standby modes in the microprocessors. An important aspect of these standby modes is not only the power consumption in each mode, but also the transition time from the standby to the active mode. A long transition time may cause a significant speed overhead and this prevents using the standby mode, resulting in high leakage-power consumption. Minimizing the leakage current for various phone operating scenarios is therefore important.

This section describes the techniques used to reduce the leakage power used in an SH-Mobile3 processor. A notable feature is the implementation of on-chip power switches, which enables two new hierarchical standby modes: resume standby (R-standby) and ultra standby (U-standby) modes.

10.2.1 Basic Concept for Low Power

A key hint to achieving low power in cellular phones is that applications that run on phones are more limited than on PCs. Integrating a dedicated computation engine and providing sufficient performance at a minimum operating frequency is an effective way of improving overall power efficiency. Accordingly, the SH-Mobile3 includes advanced CMOS technology and integration of high-performance-per-clock dedicated multiple computation engines such as a 1.8-MIPS/MHz embedded CPU core and a 6.55-ECM/MHz Java accelerator.

Lowering the operating frequency also enables the threshold voltage to be raised. The operating frequency of the SH-Mobile3 is thus successively reduced to 200 MHz, and leakage power consumption is limited to about 1% of the total power consumption. However, this leakage current is not low enough for a cellular phone in standby mode. It should be noted that the leakage budget for the application processor is about 10 μA.

A back-biasing technique is an effective way of reducing the leakage. However, this is unsuitable for high-Vth thin-tox circuits because it cannot plug the gate-tunneling leakage current and the gate-induced drain leakage (GIDL) current is not negligible. Back biasing is also less effective in advanced process technology. Power gating, i.e. using off-chip regulators to cut off the power supply to the chip externally, is another solution. The SH-Mobile2 uses this method [4], but it requires multiple power supply channels. It is also difficult to shorten the transition time from standby mode due to the large C or L components on the power line between the chip and off-chip regulators. The SH-Mobile3 therefore uses on-chip power switches.

10.2.2 Chip Overview

The SH-Mobile3 is a system-on-a-chip (SOC) device that is implemented on a low-power SOC design platform. This platform enables advanced circuit techniques, including on-chip power switches, to be used. These include thick-tox on-chip power switches (PSWs) for plugging leakage currents, μI/O for supporting multiple power domains with a wide range of conversion functions, and a low leakage data-retention RAM. (Details of the low leakage data-retention RAM are described in Section 10.3.)

Figure 10-7 shows a chip micrograph of the SH-Mobile3. A 130-nm 5-layer-Cu dual-Vth dual-tox CMOS technology is used. The supply voltage for the core is 1.2 V with 1.8 / 3.3 V for the I/O. The operating frequency is 216 MHz under the worst PVT conditions. The chip size is 7.7 x 7.6 mm². The SH-Mobile3 integrates an embedded CPU core of the Super-H family called SH-X, a DSP, 32-kB 4-way set-associative instruction and data caches, a 4-entry instruction TLB, a 64-entry unified TLB, a 16-kB local RAM (XYRAM), a 256-kB user RAM (URAM), several media-processing IPs, such as a Java, an MPEG-4 and 3D graphics accelerators, and other peripheral modules. The on-chip power switches are called PSW1 and PSW2.

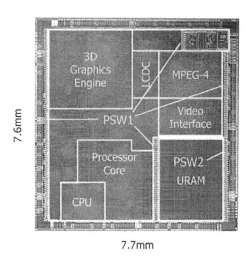

7.6mm

7.7mm

Figure 10-7. Chip micrograph of SH-Mobile3

10.2.3 On-Chip Power Switches

Figure 10-8 shows the basic configuration of the two power domains and on-chip power switches. The on-chip power switches, PSW1 and PSW2, provide local ground level (vssm1 and vssm2). PSWC is an on-chip power switch controller. The interface circuitry between the two power domains is based on a μI/O, which is described below. The critical factors in implementing the on-chip power switches are as follows:
(1) The configuration of the switches (polarity of MOS, threshold voltage, gate-oxide thickness),
(2) The size of the switches,
(3) Preventions of invalid signal transmission while the power supply on one side is cut off,
(4) Prevention of rush current when power switches are turned on.

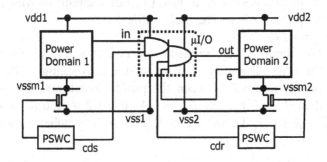

Figure 10-8. Basic configuration of two power domains and on-chip power switches

Configuration of On-Chip Power Switch

A high-Vth thick-tox NMOS transistor is used for the power switch for three reasons. First, an NMOS transistor has a g_m more than two times larger than that of a PMOS. (The on-resistance of the power switches is expressed by $1/g_m$.) Low on-resistance is essential to minimize area and speed overheads.

Secondly, there are two options for the gate-oxide thickness of the power switches: a thin-tox or a thick-tox MOS can be used in the I/O circuitry. The crucial factor is the gate-tunneling current. The gate-tunneling current in the power switches is not negligible because it flows even when the switches are turned off and it is therefore essential that they are large enough to minimize speed overheads. A thin-tox MOS has too much gate-tunneling current for the power switches to clear the leakage budget so a thick-tox MOS is the only solution possible.

Lastly, a low-Vth NMOS transistor provides low on-resistance, but it needs negative voltage to turn it off completely, thus requiring an additional power supply or on-chip voltage generator. The on-resistance of a high-Vth thin-tox NMOS may be insufficient for the power switches. However, a high-Vth thick-tox NMOS, which applies an I/O supply voltage (3.3 V) to the gates, provides sufficient on-resistance.

Size of On-Chip Power Switch

The size of the on-chip power switch should be determined so as to ensure that the speed overhead is negligible when the power switch is implemented. Large on-resistance of the power switch (R_{sw}) causes a voltage drop across the power switch, resulting in speed degradation. In the SH-Mobile3, the size of the power switches is designed so that the R_{sw} is less than 0.1% of the equivalent resistance (R_{eq}) of a circuit to which power is supplied via a power switch. The R_{eq} is defined as:

$$R_{eq} = V / I_{max},$$

where V is the supply voltage applied to the circuit, and I_{max} is the maximum current in the time resolution of the t_c of the decoupling and/or parasitic capacitors on the terminal between the power switch and the circuit. For example, when $R_{sw} = 2$ kΩ·μm, $I_{max} = 1$ A, $V = 1.2$ V, and $R_{eq} = 1.2$ Ω, the R_{sw} should be less than 12 mΩ. In total, the power switches should be more than 166-mm wide.

Prevention of Invalid Signal Transmission

It is essential to prevent invalid signal transmission between the power-on and power-off domains because this causes a significant increase in power due to short-circuit currents. The μI/O in Figure 10-8 prevents this by using a 4-input AND function. The μI/O may require a signal level-shift function when the sender's supply voltage is different from the receiver's one. Figure 10-9 shows an μI/O with a signal level-shift function. LC is the level-shifter circuitry, where a dual-rail input signal (n1 and /n1) with a signal swing of vdd1 is converted to a single-rail output signal (n2) with a signal swing of vdd2.

The 4-input AND function in the μI/O supports both internal power shutdown by the on-chip power switches and external shutdown by off-chip regulators. Three input signals (cds, cdr, and e) are used as control signals. The cds and cdr signals are automatically controlled by a power switch controller (PSWC1 or 2); the cds and cdr are driven to low when the sender and receiver domains, respectively, are turned off internally by the power switches. All the designers have to do is to drive "e" to low when the sender domain is turned off externally.

Preventing Rush Current

Turning on the power switches may cause a large rush current on power lines because large parasitic capacitances (tens of n-F) are required to charge the initial voltages. This may result in a large drop in the supply voltage. When power lines are shared with other domains, this drop in supply may be critical for other domains. Supply variation due to a supply drop can lead to timing violations. The power-switch controller implemented in the SH-Mobile3 therefore uses a slew-rate control scheme to prevent rush currents.

Figure 10-10(a) shows a block diagram of a power-switch controller (PSWC). The gate of the power switch (g) is driven at a low slew rate using two drivers (C1 and C2). Figure 10-10(b) show simulated waveforms. In the first stage (at t1), g is driven by a small driver (C1). After C3 detects that g is above the threshold voltage (Vth1) at t1a, a large driver (C2) drives g again to keep it at low impedance. A simulated rush current showed that this system reduced the rush current at a rate of over 20 dB.

The slew-rate control scheme not only prevents rush current but also provides a req/ack handshake interface so that the domain can assess the condition of the power-switch state. A timer in the power-switch controller counts from t1 to t1a (see Figure 10-10(b)), and drives ack to high at t1b, satisfying t1b = (t1a - t1)/2. Thus, when the ack signal is high, the power switch is guaranteed to be completely on. This decreases the overhead for the transition time from the standby to the active mode.

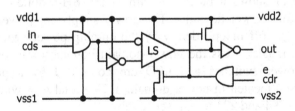

Figure 10-9. μI/O with level shifting function

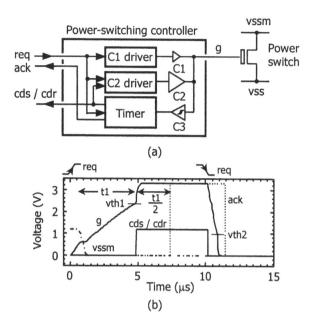

Figure 10-10. (a) Block diagram of the power-switch controller (PSWC)
and (b) simulated waveforms

10.2.4 Implementation of Power Switches

Figure 10-11 shows a block diagram of the SH-Mobile3. The chip is divided into four power domains. The power supply for power domains 1 and 2 can be cut off independently by the on-chip power switches (PSW1 and PSW2). Signals across the power-domain boundary are routed via the µI/O. The power switches and µI/O are controlled by a power-switch controller (PSC) located in power domain 3. The total gate width of PSW1 and PSW2 are 794 and 221 mm, respectively.

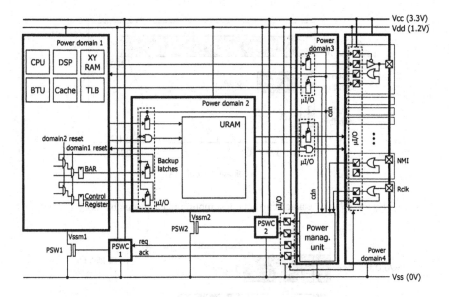

Figure 10-11. Block diagram of the SH-Mobile 3

10.2.5 Two Standby Modes

The use of on-chip power-switch technology enables two new standby modes in the SH-Mobile3: resume standby (R-standby) and ultra standby (U-standby) modes. Figure 10-12 shows the standby power consumption, or leakage current, measured at room temperature at a power supply voltage of 1.2 V. In standby mode without the power being cut off, the leakage current is 2.2 mA.

In the U-standby mode, both PSW1 and PSW2 are turned off (Figure 10-11), producing ultra-low leakage of only 11 µA. In this case, however, the transition to the active mode takes longer because most of the information on the chip is lost and the system requires a boot sequence. This standby mode is used when a flip-type cellular phone is closed.

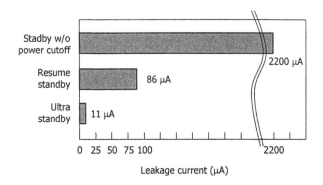

Figure 10-12. Leakage current consumption in each standby mode

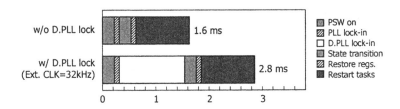

Figure 10-13. Transition time from the resume standby mode

In the R-standby mode, only PSW1 is turned off. This cuts off the power supply to the CPU core, MPEG and 3DG hardware IPs, and peripherals. The power supply to the memory and back-up registers is kept on. As a result, recovery from the R-standby mode is much more rapid than from the U-standby mode. The leakage current in this mode is 86 μA.

To ensure a quick transition from the R-standby mode with a minimum hardware cost, data backup using a backup latch (see Figure 10-11) is used. If all the information in the flip-flops is cleared by turning off the power switches, a longer recovery time is required. To avoid this, backup latches are implemented in the same power domain as the memory, and the power supply for this domain is kept on in the resume-standby mode. Key information for achieving quick recovery is stored in the backup latches before the transition to the resume standby mode, and this is restored to the original flip-flops during the recovery operation. The control-register contents that are needed immediately after wake-up, such as clock and interrupt settings, are saved to the backup latches. The boot address register (BAR) that holds the restart address is also backed up in the backup latches.

The transition time from the resume standby mode is plotted in Figure 10-13. Transition is triggered by an interrupt signal. When an interruption occurs, hardware recovery operations such as power-switch control and a

PLL lock are activated. Software recovery operations then start from the BAR address and the OS recovery routine is executed. This bar graph shows a breakdown of the recovery time. A digital PLL generates 27 MHz from a 32-kHz clock. This needs a long lock time, so if a 27-MHz clock is available for an external clock input, the recovery time is only 1.6 ms. Otherwise, the recovery time is 2.8 ms.

10.3 LEAKAGE REDUCTION IN SRAM MODULE

Application processors require three modes for actual operation. The application processor operates at high speed only when the mobile phone is processing multimedia files or games, and these periods are quite short. In this mode, SRAM is accessed every cycle. The processor also operates at low speed or is in an idle state for longer periods. The main applications are displaying the LCD screen, and processing text-based content, for example, reading or writing e-mail. In this mode, the SRAM operates at low speed, or is accessed at intervals. Low leakage current is also the key to saving power in this state. We call this mode the low-leakage active mode. Another state is the standby mode. In this mode, the SRAM is not accessed, but retains its data and the main power is consumed as leakage current. The other state is the stand-by mode.

We first consider a SRAM that fits these operating modes of the processor adaptively. We developed a 1-Mbit on-chip SRAM [6] with three operating modes: a high-speed mode, low-leakage active mode, and standby mode. Note that this SRAM features leakage-current reduction for word-drivers even in high-speed active mode. To the best of our knowledge, this is the first trial in which a leakage-reduction circuit technique has been implemented in an actual production chip to reduce leakage current in a high-speed active mode.

10.3.1 Switched-Source Control for Leakage Reduction

Below we describe the triple-operating mode of the SRAM, which corresponds to the three states of the application processors. To achieve leakage-current reduction in each mode, we introduced a switched-source control [7] scheme to the hierarchical block composition in the SRAM as a sub-divisional power-line control scheme. The 1-Mbit SRAM module is divided into four banks, and each bank is controlled by three switches driven by SW1, SW2, and SW3, respectively, as shown in Figure 10-13. SW1 drives the memory array, SW2 the word driver, and SW3 the sense amp drivers and decoders. The module also has one other controlled power line in the controller circuit, which generates memory control signals. The circuit is controlled by a switch driven by SW4. The power-switch controller generates the switch-control signals (SW1, SW2, SW3, and SW4) that are applied to each bank, thus controlling the power-line voltages individually.

In the next section, we first explain the method used to reduce the leakage current in a memory cell array.

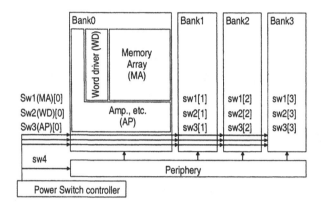

Figure 10-14. Sub-divisional power-line control scheme

10.3.2 Source-Line Self-Bias Technique

The SRAM must satisfy performance targets and is designed with sufficient stability. In this SRAM, the Vssm, the voltage of the source power line of the memory cells (Figure 10-15(a)), is increased to reduce the leakage current. This is the so-called self-reverse biasing technique [8]. As the Vssm voltage increases, the leakage current is further reduced. However, the memory cells then become less stable. To satisfy the requirements of both low leakage and high stability, the lowest Vssm that satisfies the leakage target must be generated. Therefore, we developed a new Vssm controller, the PLVC1, which consists of three nMOSFETs. One works as a power switch (MS1) between the Vssm and Vss, one as a diode (MD1), and the other as a resistor (MR1). The MR1 has a long gate and is normally on.

When a manufactured MOSFET has a high Vth, memory-cell leakage becomes smaller and the current through the MR1 is greater than the memory-cell leakage, so the Vssm voltage becomes lower. However, when the Vth is low, the Vssm voltage is high, but the MD1 restricts the rise in the Vssm voltage. This keeps the voltage low enough to retain stored data. Figure 10-15(b) shows the Vth of a manufactured MOSFET vs. the Vssm voltage. The horizontal axis shows the difference in the Vth between the designed value and the value of the manufactured MOSFET, which varies according to process variations. When the Vth of the manufactured transistor is low and the leakage current is high, the Vssm voltage has to be high to significantly reduce the leakage current. The broken line indicates the ideal values that just satisfy the leakage target and ensure the highest memory-cell stability. If the fabricated MOSFETs are worst-leakage devices, the Vth of the MOSFETs is 0.1 V lower than the designed value, and the Vssm voltage must be higher than 0.3 V to satisfy the leakage target. If the voltage controller is composed of only a power switch and a diode, the Vssm voltage is indicated by the "use only diode" line in the graph. This line satisfies the leakage target, but with low memory-cell stability.

If the voltage controller PLVC1 is composed of a constant-voltage-source circuit, the Vssm voltage is indicated by the "this work" line in the graph. This line satisfies the leakage target with more stability than when using only a diode, but the memory-cell stability is still low. Therefore, using three types of MOSFETs, MS1, MD1, and MR1, for the voltage controller, ensures that the Vssm voltage is closer to the ideal value. These values satisfy the leakage target and maintain better stability. The PLVC1 satisfies the leakage target and, at the same time, keeps the SRAM stability high.

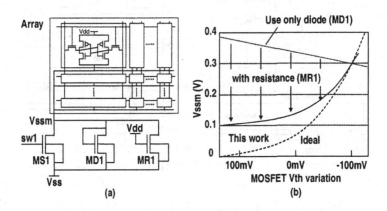

Figure 10-15. Self-biased source line voltage control

We will now examine the peripheral circuits, in particular our use of a PMOSFET switch for a word-driver with rapid recovery time. We also show how these techniques work in different modes, i.e. high-speed, low-leakage active, and standby modes.

10.3.3 Leakage-Current Reduction of Array-Associated Circuits

Figure 10-16(a) shows the leakage-current reduction scheme for an array of associated circuits that form one bank. When the memory is not being accessed, the word drivers output 0 V, and the leakage current flows through the corresponding pMOSFETs. This leakage current can be reduced by lowering the voltage of the pMOS source line, Vddw, because of the gate-source reverse biasing effect [8]. The Vddw is set to an intermediate voltage, such as 0.8 V, which achieves the leakage target and supports the quick recovery of the voltage. This is useful even in the high-speed mode as well as in the other two modes. The Vddw controller, PLVC2, is shown in Figure 10-16(a). The left pMOSFET acts as a power switch between the Vddw and Vdd. The right pMOSFET acts as a diode and restricts the Vddw potential to above 0.8 V.

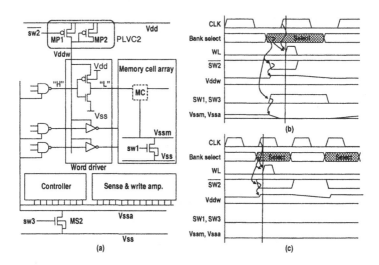

Figure 10-16. Leakage suppression in array-associated circuits

PLVC3 is composed of nMOSFETs used as power switches connected to the Vss. This reduces the leakage current of the controller and sense/write circuits in the memory array. nMOSFETs are used because they have higher conductance than pMOSFETs. The Vssa voltage is almost the same as that of the Vdd in the leakage-cut (non-active) state, but is the same as that of the Vss in the active (accessed) state.

Figure 10-16(b) shows signal transmission in the low-leakage low-speed mode. In this mode, the operating frequency is low, but the circuit operates faster because the operating conditions in the low-speed mode are the same as in the high-speed mode. This means the address signals, decoded to bank-select signals, are supplied to the SRAM module before the clock signal rises. Therefore, the bank-select signals enable the power lines to be switched to the condition of active state in time. The Vssm is discharged to the Vss level before word-line activation, and the Vssp is discharged before the assertion of the clock-signal. That is, in low-leakage mode, the state of these power lines switches every cycle.

Figure 10-16(c) shows signal transmission in the high-speed mode. The leakage control signals, SW1-SW3, are generated in response to the bank-select signals, which are generated by decoding the 2-bit address signals. The Vssm and Vddw must be in an active state, i.e., at the Vss and Vdd levels, respectively, before word-line activation. The Vssa must be in an active state, i.e., at the Vss level, before the clock rises. Since the Vddw is only connected to a word driver, it has only a low level of parasitic load. In addition, the Vddw has an intermediate voltage of about 0.8 V in a leakage-cut state. This low load and intermediate voltage enable a quick change in the Vddw voltage, so that the Vddw can be activated in time for word-line activation [8]. The Vddw can therefore switch between the leakage-cut and active states every cycle. The Vssm has a large parasitic load and it takes a long time to discharge the Vssm to the Vss level, so the Vssm cannot be activated in time for word-line activation. The Vssa also has a large parasitic load and must be at the same level as the Vss before clock assertion, which is quicker than word-line activation. This means the Vssa cannot be activated in time.

In the stand-by mode, SW1-SW3 are turned to determine the voltage in the Vssm, Vddw, and Vssa in the low-leakage state. Only data in the memory array are retained.

10.3.4 Leakage-Current Reduction of 1-Mbit SRAM Module

The remaining issue for reducing leakage current is the use of an SW4 as the controller circuits for controlling the four banks. As illustrated in Figure 10-17, which shows the whole leakage-current reduction scheme, the PLVC4 is composed of several nMOSFETs used as power switches connected to the Vss, the same as for the PLVC3. Like the Vssa, the Vssp voltage is almost the same as that of the Vdd in the leakage-cut (non-active) state, but is the same as that of the Vss in the active (accessed) state.

We manufactured a prototype chip for a 1-Mbit SRAM module using a 130-nm process. Figure 10-18 shows a micrograph of the chip. The PLVC1 power switches were made from optical dummy memory cells and were placed at the edge of the cell array. The other switches were divided and placed under the interconnects and power lines. The area penalty was therefore almost zero.

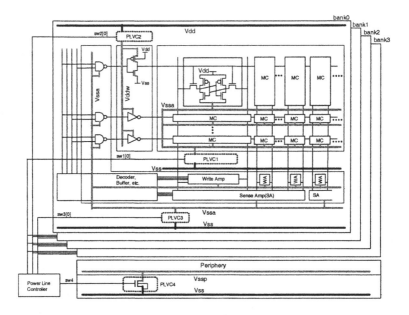

Figure 10-17. Low Leakage SRAM Module Structure

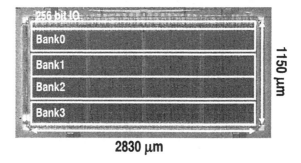

Figure 10-18. Chip micrograph of 1-Mb SRAM module

Figure 10-19 shows the measurement results for a worst-leakage sample of the prototype 1-Mbit SRAM module. The temperature was 45 degrees Celsius. The figure shows the leakage current in each operating mode. The values of conventional circuits were estimated using the results of leakage simulations and of measuring this prototype chip. The technologies described in this chapter reduced active leakage by 25% even in high-speed mode (150 MHz), by 90% in low-leakage mode (10 MHz), and by 95% in standby mode.

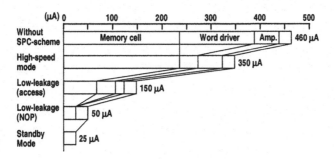

Figure 10-19. Leakage currents in 1-Mbit SRAM module

REFERENCES

[1] A. Keshavarzi, K. Roy, and C. F. Hawkins, "Intrinsic leakage in low power deep submicron CMOS ICs," *Proceedings of International Test Conference*, pp.146, 1997.

[2] H. Mizuno, K. Ishibashi, T. Shimura, T. Hattori, S. Narita, K. Shiozawa, S. Ikeda, and K. Uchiyama, "An 18-uA Standby Current 1.8-V, 200-MHz Microprocessor with Self-Substrate-Biased Data-Retention Mode," *IEEE Journal of Solid-State Circuits*, vol.34, no.11, pp. 1492, 1999.

[3] M. Miyazaki, G. Ono, and K. Ishibashi, "A 1.2-GIPS/W Microprocessor Using Speed-Adaptive Threshold-Voltage CMOS with Forward Bias," *IEEE Journal of Solid-State Circuits*, vol.37, no.2, pp. 210, 2002.

[4] T. Kamei, M. Ishikawa, T. Hiraoka, T. Irita, M. Abe, Y. Saito, Y. Tawara, H. Ide, M. Furuyama, S. Tamaki, Y. Yasu, Y. Shimazaki, M. Yamaoka, H. Mizuno, N. Irie, O. Nishii, F. Arakawa, K. Hirose, S. Yoshioka, and T. Hattori, "A Resume-Standby Application Processor for 3G Cellular Phones," *International Solid-State Circuits Conference (ISSCC), Digest of Technical Papers*, pp. 336, 2004.

[5] Y. Kanno, H. Mizuno, N. Oodaira, Y. Yasu, and K. Yanagisawa, "u I/O Architecture for 0.13-um Wide-Voltage-Range System-on-a-Package (SoP) Designs," *Symposium of VLSI Circuits Digest of Technical Papers*, pp. 168, 2002.

[6] M. Yamaoka, Y. Shinozaki, N. Maeda, Y. Shimazaki, K. Kato, S. Shimada, K. Yanagisawa, and K. Osada, "A 300MHz 25uA/Mb Leakage On-Chip SRAM Module Featuring Process-Variation Immunity and Low-Leakage-Active Mode for Mobile-Phone Application Processor," *International Solid-State Circuits Conference (ISSCC), Digest of Technical Papers*, pp. 494, 2004.

[7] M. Horiguchi, T. Sakata, and K. Itoh, "Switched-Source-Impedance CMOS Circuit for Low Standby Subthreshold Current Giga-Scale LSI's," *IEEE Journal of Solid-State Circuits*, vol. 28, no. 11, pp. 1131, 1993.

[8] T. Kawahara, M. Horiguchi, Y. Kawajiri, G. Kitsukawa, T. Kure, and M. Aoki, "Subthreshold Current Reduction for Decoded-Driver by Self-Reverse Biasing," *IEEE Journal of Solid-State Circuits*, vol. 28, no. 11, pp. 1136, 1993.

Chapter 11

CASE STUDY: LEAKAGE REDUCTION IN THE INTEL XSCALE MICROPROCESSOR

Lawrence Clark
Arizona State University, USA

11.1 INTRODUCTION

Integrated circuit (IC) power dissipation is increasingly important for all designs [1], but the need to control power is acute for hand-held devices. IC's designed for hand-held and cell phone applications require low power usage due to limited battery capacity. As the power supply voltage (V_{DD}) is reduced due to process scaling, threshold voltage (V_t) is decreased to maintain gate overdrive V_{GS} - V_t, increasing transistor sub-threshold currents exponentially [2, 3]. Standby current has become increasingly important due to this process scaling, as well as techniques such as dynamic voltage scaling (DVS) [4, 5] that limit active circuit current. Alleviating standby power using circuit methods allows use of a higher leakage process, gaining higher active circuit speed, while meeting standby power specifications. This in turn, allows higher performance or lower active power depending upon the operating voltage.

Modes of operation seemingly dominated by standby power may also have an important active component. An example is found in cellular communications, where the phone and cells must communicate on the order of once per second, so that a phone in motion can be handed off to the next appropriate cell [6]. This scenario requires an active operation interspersed in time with standby, and depending on the design, the active or standby components may dominate. It is also important that transitioning to or from a low power standby state dissipate minimal additional energy, as the expended transition energy cost must be amortized against that saved by the time in the low power mode. Difficulty in predicting, *a priori*, how long the

IC will be idle makes this more difficult. Ideally, the penalty of entering and exiting a low power state can be low enough to allow a "greedy" approach, whereby the low power mode is entered whenever possible.

11.1.1 Low Power Mode Requirements

Standby power is generally specified, at least for battery life determination, at a cool ambient (either 25°C or 50°C) since a "hand-held" device must be cool enough to hold. Battery lifetime requires IC standby currents in the 10's to 100's of μA leading to a leakage current component under 100 pA/μm of transistor gate width in the 0.18 μm technology node. Even lower leakage is required by high transistor count system on a chip (SOC) designs. This implies a V_t over 500 mV, independent of supply voltage. In fact, many "low power" processes are really *low leakage*. Arguably, these processes may be higher power where high operating frequencies are required, in that a higher V_{DD} is needed to achieve any operating frequency when compared to an otherwise identical process with a lower V_t. A low standby power mode can provide a compromise between active and standby power, providing "the best of both worlds" under some realistic operating scenarios.

To be practical, such a mode must meet some minimum necessary requirements. Firstly, the mode must provide a substantial benefit. Since the subthreshold current is exponentially related to V_t, small adjustments in V_t produce large leakage benefits and relatively small active power penalties. Consequently, any low power mode must be compared to the effect of simply changing the process threshold voltage. A mode mitigating leakage power needs to be equivalent to a V_t increase of 100 mV, i.e., delivering over 20x leakage reduction to be really worthwhile. Secondly, all designs are cost sensitive. In this regard it is important that the low leakage power mode increase the physical size of a baseline design at most modestly. Thirdly, a low standby power mode should not incur too high an active power penalty. Simply raising the V_t and V_{DD} is an alternative solution that does raise the active power, but implies very little design effort (unless V_{DD} is limited by the oxide thickness). All real IC design projects have schedules and limited design resources. A low standby power mode should therefore have limited implementation effort and risk. Very complex implementations may require multiple silicon revisions to work out unexpected bugs, and this profoundly affects the most important schedule item, i.e., shipping chips for revenue. Lastly, a circuit approach is ideally applicable for more than one process generation so that techniques and learning can be applied more than once.

11.1.2 XScale Microprocessor Project Requirements

This chapter describes the implementation of a combined reverse-body bias (RBB) and supply collapsed "Drowsy" leakage mitigation mode on the Intel XScale microprocessor [7]. Specifically, the mode was developed as part of the design effort for the 80200 microprocessor [8]. The resulting embedded core incorporated the Drowsy mode, which was subsequently used in SOC designs for handheld devices. The first XScale implementation was on a 0.18 μm process technology supporting dual gate oxides and improved metal pitches over those used on the desktop microprocessors [9] to support SOC designs. The XScale microprocessor has over 6M transistors totaling about 7 meters of transistor gate width. Separate 34kB instruction and data caches comprise 42% of the core area. The core supports operating voltages from 0.75 V to 1.5 V, with the lower values targeted at providing good performance (greater than 150 MHz) at less than 50 mW total power dissipation at 25C. It was determined that applying the resulting Drowsy scheme to the entire processor met these requirements and could achieve 100 μW standby power on a typical die.

The choice of target IC was driven by the desire to demonstrate a low standby power mode in concert with high performance, allowing the resulting embedded processor design to be placed in low power ICs operating at low V_{DD} (most likely with DVS) and high performance (non-battery powered) designs operating at high V_{DD}. The latter would most likely not utilize the Drowsy mode, but could use the core unchanged. This choice had both advantages and disadvantages. Clearly, allowing a single design to address vastly different markets minimized the overall effort. By implementing the experimental standby power circuitry in actual production ICs, large amounts of data could be collected, spanning process corners and in sufficient volume to determine manufacturability of the mode. However, the target market for the 80200 processor was not hand-held devices—other devices (such as the subsequent Intel PXA255 system on chip incorporating the same processor [10]) were intended to fill those markets. Consequently, the impact of the mode, both in design effort and on maximum performance had to be small.

11.1.3 Leakage Components

The primary leakage component exhibited by the target process core transistors was drain to source leakage current (I_{off}). The oxide thickness of 3 nm exhibited insignificant gate tunneling current. However, the approach taken here does substantially reduce gate leakage on processes that exhibit it, due to the reduction of gate bias. The baseline room temperature I_{off} of the target process was 1 to 3 nA/μm. Figure 11-1 shows the leakage current vs. drain voltage (V_{DS}) with source to bulk voltage (V_{SB}) as a parameter measured from a test wafer. If we assume that ½ of the transistors are on and ½ are off, and that only the latter contribute to leakage currents, 7 meters of gate width produces a standby current over 3 mA even at the low end of the expected I_{off}. While transistor stacking can lower the leakage as well [11] it is dependent on gate input state and inverters predominate in the design, so the overall impact of stacking is often less than 25% in our experience, as it was here.

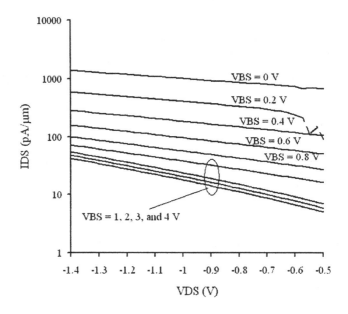

Figure 11-1. Representative 0.18 μm process PMOS I_{off} leakage vs. RBB and supply voltage.

A secondary, but important leakage effect is gate induced drain leakage (GIDL) at the gate-drain edge. It is most prevalent in the NMOS transistors, being about two orders of magnitude weaker for PMOS devices. For a gate having a 0V bias with the drain at V_{DD}, band bending in the drain region, creates electron-hole pairs. Essentially, the gate voltage attempts to invert the drain region, where holes are swept out and a deep depletion condition occurs [11]. GIDL may consequently limit the magnitude of reverse body bias that can be applied for leakage control. Eventually most of the current contribution is due to GIDL and increases faster with body bias than the I_{off} is decreased by it. The onset of this mechanism can be lessened by limiting the magnitude of drain to gate voltage used.

Diode area components from both the source-drain diodes and the well diodes are negligible with respect to I_{off} and GIDL components on the target process. However, on 90 nm and smaller processes, steep doping gradients at the drain to bulk interface are making direct band-to-band tunneling a significant leakage current contributor. In order to apply leakage reduction schemes such as described in this chapter on these scaled processes, the transistor design may need to comprehend and limit this component.

11.2 CIRCUIT CONFIGURATION AND OPERATION

11.2.1 Design Choices

Multi-threshold CMOS (MTCMOS) connects the logic circuit to a virtual power supply node that is gated through a high V_t transistor, rather than connecting the circuit directly to the power supply itself [12]. When the supply gating transistor is in cutoff, the virtual supply collapses to the other supply due to core circuit leakage. The resulting standby power is due entirely to leakage through the gating transistor. Since this transistor can be relatively small and has a higher V_t, very low standby power can be obtained. Unfortunately, collapsing the core voltage completely in this manner forces the logical state of the IC to be lost. Such non-state retentive "sleep" modes incur power penalties: The present logical state must be saved before sleep and restored before resuming active operation, requiring a separate low standby power storage medium; The data movement increases latency; The I/O power must be amortized by the static power savings achieved during the time in sleep, precluding frequent use. These considerations led to the choice of a state retentive approach. A state retentive RBB implementation was described in the previous chapter. Unfortunately, when this approach is used on a 0.18 μm or smaller process, the high voltages exacerbate GIDL and drain edge tunneling leakages. A different scheme that limits voltages while applying RBB is required.

Initially, circuit configurations that minimized design effort were investigated. These are illustrated in Figure 11-2. The circuit of Figure 11-2(a) applies the body bias to both the NMOS and PMOS via diodes. This has a number of disadvantages. First, diodes don't supply much current until the bias is greater than 0.7 V and since two forward diode voltages must be supported the minimum supply must be nearly 2V, assuming the core collapse would be limited to about 0.5 V to retain state. Secondly, the active power must be supplied to the core via transistors M3 and M4, which incurs two resistive drops through gating transistors. Figure 11-2(b) uses diode-connected transistors to provide a smaller voltage collapse. Here, the NMOS V_t sets the PMOS body bias and vice versa. Since these track independently, there is no guarantee of reaching even an arguably optimal standby mode operating bias.

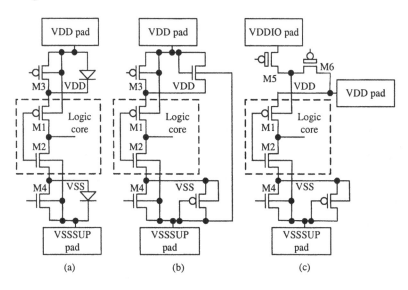

Figure 11-2. Possible RBB circuit configurations that provide simultaneous core voltage collapse. The box symbols indicate thick gate high voltage tolerant transistors.

Minimizing the active power drop due to gating transistors leads to the configuration shown in Figure 11-2(c) that incurs this drop on only one power supply, in this case V_{SS}. RBB can be applied to the PMOS by pulling the N wells to V_{DDIO} (recall PMOS transistors have limited GIDL) via thick gate transistor M5. Transistor M6 shorts V_{DD} and the well voltage during active operation. An NMOS gating transistor provides higher drive strength, especially if a thicker gate, higher V_t transistor such as that intended for input/output (I/O) circuits is used, as V_{GS} can be V_{DDIO}, which was 3.3V on the target process here. By tying the substrate to V_{SSSUP}, i.e., ground, on this N well process, RBB is applied to the NMOS transistors as V_{SS} rises. A number of simple configurations were investigated to allow the V_{SS} node to rise in Drowsy mode but still provide enough current to avoid $V_{DD} - V_{SS}$ from collapsing too far, that is, far enough to allow logic state loss. It was found that diodes and diode connected exhibited sufficient variation across process corners that only a small leakage reduction could be counted on. An IC specification must be worst case, leading to the conclusion that some form of regulation would be required when applying the body biases for consistent and near optimal leakage reduction.

11.2.2 Drowsy Mode Circuit and Operation

The circuit configuration used is shown in Figure 11-3. This mode utilizes RBB as well as V_{DD} - V_{SS} collapse to limit leakage power [13], achieved via large supply gating transistors that allow the NMOS source to be raised as mentioned. They also allow full collapse of the core voltage, which produces the non-state retentive "sleep" mode, essentially the classical MTCMOS approach to leakage control at no cost. Drowsy mode is fully state retentive and is exited on any interrupt, while sleep mode is terminated by asserting reset—since state is not retained, returning from sleep mode is essentially a cold start. Power is applied to the V_{DD} and V_{SSSUP} pins. As mentioned, power supply IR drop is avoided by having only one power carrying supply gated by a series transistor. Large N channel transistors M5 provide V_{SS} to the active circuitry during normal operation. Large P channel transistors M7 provide clamping of the N well (V_{DDSUP}) to V_{DD} to maintain 0 V_{SB} on the core PMOS transistors while in active operation. These transistors must have an oxide thickness suitable to exposure to high voltages as indicated by the box gate symbol. The NMOS clamp transistors are less than 2% of the total transistor width of the microprocessor, while being over 20 times the total drive of the PMOS clamping transistors.

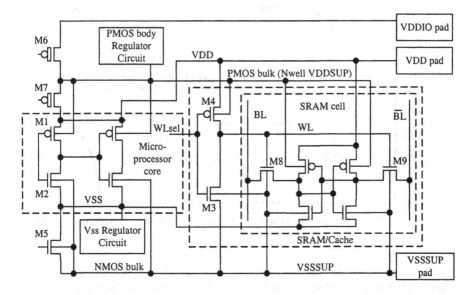

Figure 11-3. Drowsy mode circuit configuration for core and SRAM.

Figure 11-3 also shows that a negative gate to source voltage is applied to SRAM access transistors M8 and M9 via word-line pull down transistor M3, which has a source connection to V_{SSSUP}. This negative gate to source voltage essentially eliminates the I_{off} of those transistors that would otherwise comprise 42% of the SRAM leakage as determined through circuit simulation [14]. This technique was also adopted for an SRAM on a low leakage 130 nm process [15]. We have found that on a 130 nm process with high gate leakage [16] the added gate current produced by the 0.65 V across the oxide was greater than the drain to source leakage savings. Thorough understanding of the target process behavior is essential when choosing low standby power techniques.

For sleep as well as Drowsy modes, the M5 clamp transistors are in cutoff. In the former case, the core V_{SS} is allowed to float up towards V_{DD}, as shown in Fig 11-4. In the non-state retentive sleep mode, V_{SS} can nearly reach V_{DD} and power consumption is limited to the leakage current through the NMOS clamp devices. The clamp devices should be high V_t as in MTCMOS, to minimize this current since they do not have body bias applied. The higher V_t thick gate devices intended for IO are adequate. That said, in order to avoid compromising active performance in the 80200 implementation, the V_{SS} clamping transistors in this first design used relatively leaky (lower V_t) thin oxide core transistors.

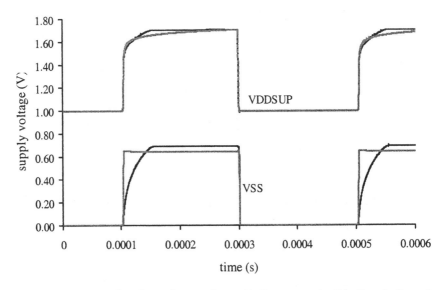

Figure 11-4. Power supply voltages in operation and in Drowsy mode. Grey lines indicate the operation at 100°C while black lines indicate 25°C. The V_{DD} and V_{SSSUP} stay at 1 V and 0 V respectively and are not shown.

In Drowsy mode, the rising V_{SS} applies RBB to the NMOS devices, but it is regulated to avoid losing state from too little V_{DD} - V_{SS} (gate overdrive) voltage. This regulation is apparent in the supply waveforms in Figure 11-4, at both 25°C and 100°C where the V_{DD} - V_{SS} is about 350 mV. Note that the logic circuits are operating in subthreshold at this bias condition. However, the goal is to retain state. Circuits in Drowsy mode are not allowed to toggle state. Wakeup circuitry is clocked by the PLL reference clock and does not enter Drowsy. Raising the NMOS source voltage rather than decreasing the NMOS body voltage is advantageous since it does not require a twin tub or triple well process or charge pump circuitry. Additional decrease in I_{off} is derived from lower V_{DS} (corresponding to the DIBL component of V_t) as well as limiting GIDL components as mentioned in section 11.3.1. Referring to Figure 11-1, the RBB accounts for about 10x leakage reduction and the collapsed V_{DS} for the remainder. Since gate current is strongly affected by the gate to source voltage [17], the lower voltages across the gate oxide can afford substantial decrease in that component on processes with thin oxides.

The N well voltage is raised to reverse body bias the PMOS transistors at the same time as V_{SS} rises as shown in Figure 11-4. There is no deleterious effect from using a large bias on the well as long as circuit configurations that accumulate the gates of the PMOS transistors are avoided. However, changing the voltage of this capacitance constitutes an overhead power. The current reduction due to increasing RBB is limited since the body effect follows a square-root dependency. Consequently to keep the bias optimized, two methods were supported: For low I/O voltages, i.e., 1.8 V, the well can be directly connected to the I/O supply during Drowsy. If the I/O voltage is 2.5 or 3.3 V, a regulator is used.

11.2.3 Power Grid and Layout Considerations

To apply RBB, separate bulk and source supplies are required for both P and N transistors. Nonetheless, the area penalty to support RBB can be very small. The power supply clamping transistors (M5 and M7 in Figure 11-3) are provided in the pad ring only, occupying otherwise empty or I/O decoupling capacitor space within the supply pins. Since they are not required to provide high currents, the bulk connections are routed sparsely through the logic circuitry, limiting the density impact. A two layer routing grid with 50 μm between body supplies is utilized, as shown in Figure 11-5. For auto placed and routed (APR) blocks, taps were inserted during placement and short routes to the sparse V_{SSSUP} and V_{DDSUP} lines minimize the impact on routing density as evident in the figure. V_{SS} and V_{DD} are routed through each cell row in the conventional manner. No long horizontal tracks are wasted on the extra supplies at the cell level. Thus, the design work required to support RBB was also small. Taps were removed from the standard cells, new tap cells were inserted into the library, and SRAM tap cells were enlarged to allow V_{SSSUP} and V_{DDSUP} routing. The substrate is highly doped below an epitaxial layer, providing an effective short circuit between V_{SSSUP} (ground) rails and limiting noise due to switching. N-wells are intentionally contiguous to grid at the substrate level for V_{DDSUP}. This was enforced through all custom and APR blocks.

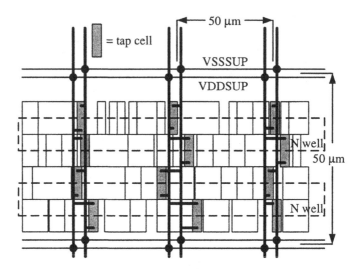

Figure 11-5. Power supply routing grids

Referring back to Figure 11-3, the body connections of transistors M6 and M7 are to V_{DDIO}. The PMOS clamping transistors carry no DC current and are 17 mm in total width. The NMOS clamp transistors are 85 mm in total width to deliver the entire active power of the core. Both the NMOS and PMOS power supply clamps were located only in the chip pad ring at 17 power and ground pads each. A large 55 nF on-die decoupling capacitance avoids a large instantaneous voltage drop across the NMOSFETs in active operation. This allows small clamp transistors since the clamps need only supply an average current and not an instantaneous one. This is important since arguably, it may not be possible to supply high instantaneous currents unless the clamp transistors are located throughout the design with fine granularity and very large total size. The decoupling capacitance was inserted opportunistically between circuit blocks and helps high frequency operation as well as Drowsy mode.

11.3 REGULATOR DESIGN

11.3.1 V_{SS} Regulation

The V_{SS} regulator circuit is illustrated schematically in Figure 11-6. The regulator is a three-stage amplifier with an NMOS output transistor M16. Three stages are needed due to the low operating current required to keep the regulator power consumption a negligible contributor to the total standby power. At the typical process corner with a 1 V V_{DD} it consumes about 4 μA, excluding the output transistor. The input stage amplifier includes the differential pair transistors M12 - M15 plus the bias stack and compares the voltage on V_{SS} with a reference voltage. The reference voltage is generated by PMOS transistors MR0 – MRn simulating a resistor stack, which allows it to vary with power supply. The body connections keep the biases of all transistors in the "resistor" stack identical, closely mimicking linear resistors. This is simpler than say, a band-gap reference and allowed greater V_{DD} - V_{SS} while simultaneously increasing RBB by raising V_{DD}. Since each PMOS transistor is biased in subthreshold, the stack current is under 100 nA and is left continuously connected in all modes.

The output transistor M16 can only pull V_{SS} towards 0 V, so V_{SS} rises passively, driven by the core leakage. Power would not be saved by actively raising V_{SS} and consequently, no power is wasted on circuitry to drive V_{SS} towards V_{DD}. In the event that the Drowsy mode is prematurely terminated, e.g., to service an interrupt before V_{SS} reaches equilibrium, the energy required to charge and discharge this highly capacitive power supply node is saved. The output transistor is sized to provide the full IC leakage current at high temperature and the worst-case process corner. The second buffer stage provides increased voltage output range and current drive to the gate of M16. The regulator amplifier slew rate and gain are low due to the low bias current levels, which makes the step response poor. To address this, the buffer stage devices includes the diode-connected transistor M11 that, combined with the sizes chosen, delivers the DC characteristic shown in Figure 11-7. This keeps transistor M16 from completely cutting off and limits the required slew rate of the buffer stage driving the large capacitive load presented by M16.

The entire circuit, amplifier and the IC leakage must comprise a stable system so adequate phase margin must be maintained. This is particularly important since even a momentary overshoot on V_{SS} may cause loss of state. Essentially, the IC leakage presents the load current for output transistor M16 with the AC equivalent circuit a rather complicated function of the body transconductance presented by the core NMOS transistors, the decoupling capacitance due to V_{SS} capacitance and intentional decoupling devices, an "auxiliary" RC due to the conducting NMOS transistors and the gates on the far side of those transistors, as well as the amplifier nodes. Fortunately, the circuit poles may be approximated by the dominant terms, greatly simplifying the analysis. For the V_{SS} node, this comprises just the output conductance of transistor M16, while the amplifier dominant pole is the output of the differential pair. The latter pole is the first, at approximately 65 kHz calculated from the small signal parameters. The simulated Bode plot for V_{SS} is shown in Figure 11-8. The low gain of the amplifier is advantageous, leading to a low unity gain bandwidth and greater than 60 degrees of phase margin at all process corners. The highly capacitive, low pass nature of the V_{SS} does not require high amplifier speed to stabilize. Slow turn on and turn off of the supply gating transistors (M5 in Figure 11-3), achieved by under-driving the gate of M5 also helps avoid overshoot and inductive effects on the V_{SS} supply node.

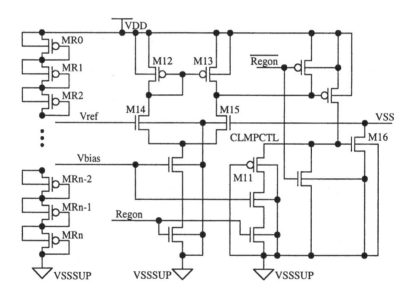

Figure 11-6. The VSS regulation circuit

11.3.2 PMOS Bulk Regulator

The regulator driving the N well V_{DDSUP} node connected to the PMOS bodies utilizes a bootstrapped voltage reference driving a wide NMOS vertical drain transistor in a source follower configuration as shown in Figure 11-9. This transistor (M1 in the figure) can provide the well and diode leakage current while operating in the subthreshold region of operation, which results in a low drop from the reference voltage to V_{DDSUP}. In fact, the drop is slightly negative at most process corners, as confirmed on silicon. The vertical drain configuration allows this thin gate oxide device to tolerate high drain to gate voltages by using an N well as the drain of a non-self aligned transistor.

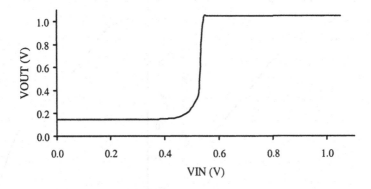

Figure 11-7. Buffer stage voltage input vs output characteristic.

As evident in the figure, this circuit operates in open loop, setting V_{REF} to approximately one thick gate threshold voltage above the V_{DD} power rail. The circuit operates between the V_{DD} and V_{DDSUP} power rails, allowing V_{DD} to be the lower voltage reference and forcing the N well RBB voltage to track upwards with it. The large V_{SB} on the thick gate NMOS transistors M6 – M9 increases the V_{REF} voltage as well. At V_{DDSUP} below the nominal output voltage, the circuit delivers V_{DDSUP}. The reset circuit (transistors M3 and M4) is required since in any bootstrapped circuit, the zero current state is a valid state, but must be avoided. Hence, the V_{REF} node is pulled up during reset to force the bootstrapped circuit out of that operating point. VBIAS is set to limit current through MVB, simulating a high value resistor. The stacked configuration limits leakage to V_{REF}.

11.3.3 Measured Results and Design for Test

The elimination of the requirement for external state memory is one of the primary attractions of the Drowsy mode, so it is essential that state retention be verified. Pre-charged (domino) circuit paths (which also imply transparent latches) exist, as do memory elements in the form of flip-flops, register files, and SRAM's. All must be tested to establish a failure due to excessive voltage collapse, especially since defects may affect the fail point.

The Drowsy mode described here presents a number of circuit debug and validation challenges [18]. These include verifying state retention, determining and setting optimal voltage regulator operating points and margin in manufacturing, and eliminating noise induced upsets and incorrectly connected circuits.

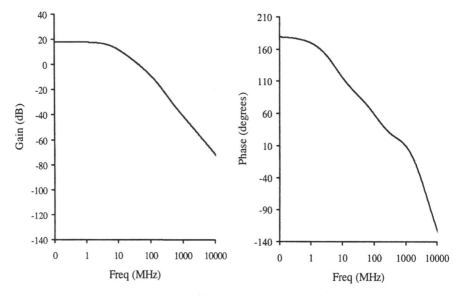

Figure 11-8. Plots of the V_{SS} regulator gain and phase margin.

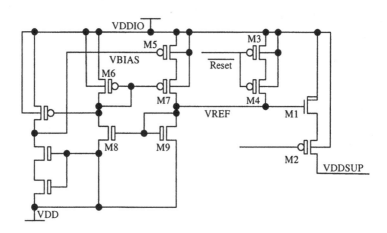

Figure 11-9. PMOS body-bias regulator driving the core N well voltage V_{DDSUP}. All NMOS bulk connections are to V_{SSSUP}.

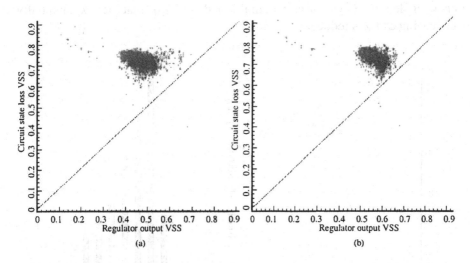

Figure 11-10. Measured state loss V_{SS} vs. regulator output V_{SS} with (a) and without a 100 μA guardband current added during test (b).

The voltages applied to produce RBB are critical, given the limited V_{DD} voltage, in this case 1 V, for successful operation. The V_{SS} must optimize the RBB to the NMOS transistors while limiting other currents by collapsing V_{DD} - V_{SS} as far as possible without state loss. Consequently the bias regulators, particularly the regulator controlling V_{SS}, are important to characterize and test. To this end, separate pad connections to the V_{SS} and N well (V_{DDSUP}) provide both observability and controllability in the 80200 design. These allow the operation of the regulators to be measured "in-situ" by using them as outputs. The failing voltage, i.e., where state is lost, is accomplished by disabling the regulators and driving test voltages into the IC. Additionally, and critical for manufacturability, they afford the ability to source additional current to determine amplifier margin, i.e., the amount of guardband in the regulator I-V characteristic, on each die during operation. Results of these measurements across multiple wafers are shown in Figure 11-10(a) which shows the circuit fail point, i.e., the worst case state loss V_{SS} as determined by a state failure across multiple tests vs. the regulator output V_{SS} as measured on that particular die. The data is taken across process corners on silicon. Clearly, ICs measuring below the diagonal line are failing. There is a substantial reduction factor in the leakage current in Drowsy mode. Figure 11-11(a) shows the current with the PLL off (no clocks and no body bias applied) compared to the Drowsy mode, where I_{off} is reduced by over 25x. Figure 11-11(b) shows the distribution of the Drowsy current across one wafer. The shape of the distribution is due to the Gaussian channel length (L_E) variation multiplied by I_{off} that is exponentially increasing as L_E is reduced.

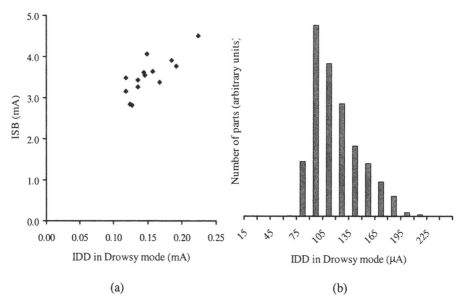

Figure 11-11. Measured I_{SB} vs. Drowsy I_{DD} current (a) and distribution of Drowsy I_{DD} currents over one wafer (b).

Figure 11-10(b) shows the shift due to 100 µA added current forced into the V_{SS} node, nominally doubling the standby current. The effect, essentially shifting the data to the right and down is apparent. In this manner, die with nominally working Drowsy modes, but with poor margin can be rejected. Typically, V_{SS} and N well pads implemented to facilitate test and characterization are made inaccessible to the end user to avoid interfering with normal operation. This is accomplished by fusing access away after test or by employing a production package that does not provide access. In this latter case, debug requires a separate package with the needed pad connectivity (i.e., a bond option). Of course, this approach requires that regulator measurement and calibration occur at wafer sort rather than packaged product testing. Systematic as well as random process variation are increasing with each process generation. Due to these variations the leakage, fail point, and optimum regulator set point may vary from die to die. The reference voltage can be programmed via fuses or by other non-volatile on-die memory. By measuring the device fail point, determining the current load placed on the regulator, and measuring the regulator with added current for margin, the set point can be programmed to maximize RBB efficacy while maintaining operating margin.

11.4 TIME-DIVISION MULTIPLEXED OPERATION

This section will show that time division multiplexing (TDM) between active operation and a state retentive low power mode can effectively simulate a low leakage process [19]. This allows the choice of a leakier, higher performance process affording high frequency operation if needed while still meeting standby power requirements. As mentioned above, it is at low frequency operation where leakage predominates over active power. The objective of TDM operation is to run at a high enough clock rate, e.g., 300 MHz, whereby there is sufficient time between computations to limit the leakage using the Drowsy mode, even though the "effective frequency" (F_{EFF}) as measured by the amount of computation done in one second may be much lower, say 15 MHz. In this example, 95% of the time can be spent in the Drowsy mode. This is in contrast to conventional designs where the clock is slowed to 15 MHz.

The power of the IC has different dominant components in the two primary operating modes. In active mode, the active switching power plus the active mode leakage constitute the total power. In Drowsy mode, the reduced leakage and the regulator power dominate. The previous section showed that the regulation power is very small, on the order of 5% of the Drowsy mode power. The PLL lock operation and power supply movement contribute to the penalty power to enter and leave the low standby power mode. The total energy cost of a single entry into the low power mode is then 30.6 nJ. This is equivalent to about 60 active clock cycles on the processor when operating at a 1 V V_{DD}. However, the leakage savings must amortize this expended energy, so to have benefit the IC must remain in Drowsy much longer than that.

11.4.1 Apparatus

These experiments were all performed on a system at 1V V_{DD} running IO at 100 MHz and with a core frequency of 333 MHz. The board was modified to separate the microprocessor core and analog supplies from the on-board power supply, allowing external power supply and ammeter connections. The 80200 microprocessor was tested on the board to 800 MHz at 1.55 V. Hence the inclusion of the drowsy mode did not seriously impact maximum performance (at the time of this writing, the 80200 is still the fastest production ARM compatible microprocessor). A PC recorded the measured supply current and programmed the interrupt frequency (FD) driven by an on-board timer. The target voltage for the core was adjusted based on the reading from the voltmeter in order to account for IR loss in the power supply leads. Another PC controls the program running on the 80200. A representative power measurement is shown in Figure 11-12.

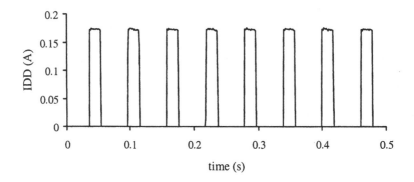

Figure 11-12. Power supply current for a representative measurement.

The program running on the microprocessor core was intended to be similar to the workload seen in an interactive application, where user input and the resulting computation occurs rarely (less than 100 Hz). The amount of computation will vary by application, so this was a parameter in the program. The resulting code is similar to the following nested loop:

```
outerLoop:
        MOV   R0, #instructions_per_interrupt
work:  SUBS  R0, R0, #1   ; decrement count
        BNE   work         ; loop while count != 0
        DROWSE             ; wait for interrupt
        B     outerLoop
```

The number of instructions to run at each interrupt is set by the parameter loaded by the first instruction. At the end of the inner loop the IC returns to the low standby power mode. Interrupts end the low standby power mode and begin the loop anew. Due to branch prediction, the processor executes one instruction per clock in this loop. State retention while in Drowsy mode maintains the cached instructions.

11.4.2 Experimental Results

The standby current in Drowsy mode was measured to be 0.1 mA at 1 V on the IC used in these measurements. The use of multiple parts was precluded by the need to solder the device to the measurement board, but the power in the Drowsy mode was typical of that measured for other devices as shown in Section 11.3.3. The separate analog PLL and clock divider supply allows independent measurement of the PLL power as well as the IC non-RBB standby power. For the IC used, in a DC condition, the room temperature core I_{DD} with no clock running was 2.8 mA at 1 V. The PLL consumed 6.6 mW at the same V_{DD}.

The processor was run at a number of interrupt frequencies and instruction per interrupt rates. As expected, the power shows a linear dependency on F_{EFF} at high rates, where the active power dominates, while at low F_{EFF} there is a floor due to leakage power (see Figure 11-13). All interrupt and instruction rates fall on the same curve as shown, demonstrating a low penalty power. Figure 11-13 shows the power vs F_{EFF} when the PLL is kept active, no RBB is applied, but clocks are gated at the PLL (labeled idle) as well as when the PLL is not running (labeled ISB) and using Drowsy mode. The power savings of Drowsy over idle mode is nearly 100x and it still obtains 28x power reduction at low F_{EFF} when the PLL is disabled. The current in the latter mode corresponds to the I_{SB} of the part.

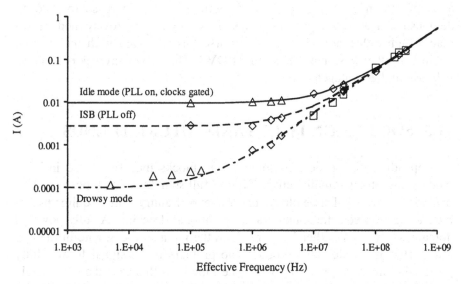

Figure 11-13. Measured power supply current vs. effective frequency with three low standby power modes.

11.4.3 Effect on Active Power

The active power domination at high operating frequencies does not imply that devices capable of high operating frequencies do not need low standby power modes. Quite the opposite, a lower threshold voltage allows higher performance at lower power for all frequencies, but the use of the TDM between active and Drowsy modes effectively simulates low leakage, when low frequencies of operation are appropriate to the required computational demands of the moment. This is likely to be most of the time for a hand-held device, but does not preclude higher peak performance, demand for which is consistently increasing with time. A lower V_t implies higher leakage power, mitigated here by the low power mode, but also higher performance at low voltages. The lower active power, resulting in higher full V_{DD} performance, also allows higher performance at low V_{DD}. The active power reduction factor can be estimated by considering the V_t increase required to match the I_{off} reduction and the required V_{DD} increase to achieve the same performance. Increasing V_t by 110 mV to 500 mV results in the same reduction. At this higher value of V_t, the frequency obtained with a V_t of 390 mV at $V_{DD} = 0.75$ V requires a V_{DD} increase to 0.86 V, demonstrating an active power savings of 24% by using Drowsy mode rather than a higher threshold voltage. Of course this will vary with the specific circuits in other ICs, and this assumes DVS. The power savings percentage is less at higher voltages.

11.5 SOC DESIGN ISSUES AND FUTURE TRENDS

Depending on the design, some circuit blocks may be placed in RBB mode while others remain active. Even in full standby some blocks must be operational, e.g., real-time clocks and wakeup circuitry. This requires proper interfacing between the standby and operational domains. A fully powered domain will present either V_{DD} or 0 V at the input of a cell gate in the low power standby mode. This produces no problem if the signal from a fully powered domain drives a transistor gate input—In this case the 0 V merely applies a negative gate to source voltage. However, if the cell input is a CMOS transmission gate or other pass gate configuration, i.e., drives a diffusion or pass gate input, care is required. When a 0 V input will be more than V_t below the gate voltage and an otherwise "off" pass-gate will allow a logic 0 to write the other side. A latch circuit will fail with logic 0 inputs, just as when coupled noise drives a latch pass gate input to a voltage more than a threshold voltage below the V_{SS} supply rail. A multiplexer (mux) circuit can create a DC current path as illustrated in Figure 11-14. The mux drivers are in a fully powered domain and drive logic 0 as 0 V to the mux inputs controlled by a logic 0 at 650 mV. Forbidding pass gate inputs at the body-biased domain interface avoids this problem. Inputs from the voltage collapsed (body-biased) domain to the operational (full voltage) domain are another matter. Here, any logic 0, represented by approximately 650 mV, will appear as a mid-rail input (a poor logic 1) and cause a DC current in the next gate. To avoid this, all interfaces passing signals from operational to body-biased domains must have latches that are closed before the inputs enter the standby mode, thus capturing the last valid state. Clearly, the latches must be in the fully powered domain.

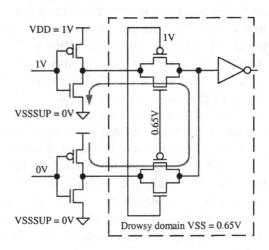

Figure 11-14. Multiplexer circuit illustrating the sneak path current created when driving pass gate inputs from the wrong (non-Drowsy) domain.

The design presented here was implemented in a 0.18 μm technology, where I_{off} was the dominant leakage contributor. Oxide scaling is dramatically increasing I_{gate} at the sub 130 nm technology nodes and as mentioned, the Drowsy biasing, by applying RBB simultaneous with supply collapse is effective in limiting I_{gate}. An implementation on a 130 nm SRAM demonstrated 16.7 pA leakage per SRAM cell [15] and use on a 65 nm process aimed at hand-held ICs has also been suggested [20]. Both leakage and SOC transistor counts are increasing so maintaining the same standby current over time requires new schemes that are increasingly effective. An obvious step is to use a scheme like the Drowsy mode presented here only on the state retaining, i.e., latch and memory, circuits while fully collapsing the supply on the rest [20]. This affords about 3x greater reduction, based on average IC's having about 1/3 of their transistor width in such memory elements. Other schemes, relying on high V_t [21] or thick gate shadow latches [22] are much more expensive in terms of size impact but promise even greater leakage reductions. They are however, difficult to apply to SRAM, where circuit density is most important. Use of thicker oxide in SRAMs [23], while difficult and costly, may eventually be required.

11.6 CONCLUSION

The implementation of a low power Drowsy mode employing RBB and supply collapse to suppress leakage currents on the XScale microprocessor has been described. The results show effective simulation of a low leakage process using Drowsy mode on a high performance 0.18 μm CMOS process. Low standby power of 100 μW and high efficiency is achieved with regulation costing on the order of 5%. The implementation effort was shown to be small and the approach applies to SRAM circuits without additional area penalty. Critical to practical application of such a low standby power mode, the cost of entry and exit is approximately the same as executing 60 instructions. Compared to the conventional low latency approach of gating clocks after the PLL, nearly 100x power savings was demonstrated. Latency in leaving the low standby power mode is limited by the PLL lock time, which ranges from two to 20 μs. This latency was fixed on the design presented here, but can be eliminated by running the IC at a low rate (at the PLL reference clock) until the PLL is locked. This is in contrast to present non-state retentive sleep modes that have much greater latency, depending on cache state, and power penalties that are much larger.

The emergence of processors supporting DVS is due to the wide range of operational frequencies that hand-held devices must support, while keeping the energy consumed as low as possible. For instance, the processor here reaches peak performance of 800 MHz assuming the operating voltage is raised to meet such a computing demand, while still meeting 100 μA standby power. The Drowsy mode allows threshold voltage to be lowered without sacrificing the power consumed at low, or even vanishing frequencies. Simultaneously, it allows lower active power at the same performance levels by allowing lower V_{DD}. TDM operation, simulating low leakage, can be applied to future designs incorporating supply collapse, RBB, MTCMOS, or combinations thereof.

Limitations of this Drowsy mode implementation stem from worst-case mismatch between the state retaining on devices and the attached leaking transistors. This is worst-case for high fan-in domino circuits and highly imbalanced latches. Future work could press these voltages lower by using MTCMOS and only well-balanced latches using RBB and supply collapse, resulting in greater than six-fold improvement in transistor counts or standby power.

REFERENCES

[1] T. Mudge, "Power: A first-class architectural design constraint," *Computer*, 34(4), 52-57, (2001).

[2] T. Burd, T. Pering, A. Stratakos, and R. Broderson, "A dynamic voltage scaled microprocessor system," *IEEE J. Solid-State Circuits*, 35(11), 1571-1580, (2000).

[3] L. Clark, et al., "An embedded 32b microprocessor core for low-power and high-performance applications," *IEEE J. Solid-State Circuits*, 36(11), 1599-1608 (2001).

[4] Y. Taur and T. Ning, *Fundamentals of Modern VLSI Devices* (Cambridge University Press, Cambridge, UK, 1998).

[5] N. Kim, et al., "Leakage current: Moore's law meets static power," *Computer*, 36(12), 68-75 (2003).

[6] H. Holma and A. Toskala, eds., *WCDMA for UMTS: Radio Access for Third Generation Mobile Communications*, (John Wiley and Sons, NY, 2001).

[7] M. Morrow, "Microarchitecture uses a low power core," *Computer*, 36(4), 55 (April, 2001).

[8] Intel® 80200 Processor based on Intel® XScale™ Microarchitecture Data sheet. Available at http://www.intel.com.

[9] M. Bohr, et al., "A high performance 180 nm generation logic technology," *IEDM Tech. Dig.*, 197-200 (1998).

[10] Intel® PXA255 Processor Data sheet; http://www.intel.com.

[11] S. Wolf, Silicon Processing for the VLSI Era: Volume 3 – The Submicron MOSFET, *Lattice Press*, Sunset Beach, CA, 1995.

[12] S. Mutoh, T. Douseki, Y. Matsuya, S. Shigematsu, and J. Yamada, "1-V power supply high-speed digital circuit technology with multithreshold-voltage CMOS," *IEEE J. of Solid-state Circuits*, 30(6), 847-854 (1995).

[13] L. Clark, N. Deutscher, F. Ricci, and S. Demmons, "Standby power management for a 0.18 μm microprocessor," *Proc. Int. Symp. Low Power Electronics and Design*, 7-12 (2002).

[14] B. McDaniel and L. Clark, *U.S. Patent 6,166,985*: "Integrated circuit low leakage power circuitry for use with an advanced CMOS process," (2000).

[15] K. Osada, Y. Saitoh, E. Ibe, and K. Ishibashi, "16.7 pA/cell tunnel-leakage-suppressed 16Mb SRAM for handling cosmic-ray-induced multi-errors," *Proc. Int. Solid-state Circuits Conf.*, 302-304 (2003).

[16] S. Tyagi, et al. "A 130 nm generation logic technology featuring 70 nm transistors dual Vt transistors and 6 layers of Cu interconnects," *IEDM Tech. Dig.*, 567-570 (2000).

[17] R. Krishnamurthy, A. Alvandpour, V. De, and S. Borkar, "High-performance and low-power challenges for sub-70nm microprocessor circuits," *IEEE Custom Int. Circuits Conf. Proc.*, 125-128 (2002).

[18] L. Clark, D. McCarroll, and E. Bawolek, "Characterization and debug of reverse body bias low power modes," *Electronic Device Failure Analysis*, 6(1), 13-21 (2004).

[19] L. Clark, M. Morrow, and W. Brown, "Reverse Body Bias and Supply Collapse for Low Effective Standby Power," *IEEE Trans. on VLSI Systems*, 12(9), 947-956 (2004).

[20] S. Zhao, et al., "Transistor optimization for leakage power management in a 65 nm CMOS technology for wireless and mobile applications," *VLSI Symp. Tech. Dig.*, 1-15 (2004).

[21] S. Shigematsu, S. Mutoh, Y. Matsuya, Y. Tanabe, and J. Yamada, "A 1-V High-speed MTCMOS Circuit Scheme for Power-Down Application Circuits," *IEEE Journal of Solid-state Circuits*, 32 (6), 861-870 (1997).

[22] L. Clark and F. Ricci, *US Patent #6,639,827*: "Low Standby Power using Shadow Storage," 10/28/03.

[23] L. Clark, R. Patel, and T. Beatty, "Managing Standby and Active Mode Leakage Power in Deep Sub-micron Design," *Proc. Int. Symp. Low Power Electronics and Design*, 274-279 (2004).

Chapter 12

TRANSISTOR DESIGN TO REDUCE LEAKAGE

Sagar Suthram, Siva Narendra[§] and Scott Thompson
Univeristy of Florida, USA and [§]Tyfone, Inc., USA

12.1 INTRODUCTION

Planar Silicon (Si) MOSFETs were once promised as a device technology that dissipated negligible power in the off-state. However, in the pursuit of high performance, the channel length has been aggressively scaled increasing the nanoscale transistor leakage to near the practical limit of ~100 nA/um. The high off-state leakage is an indication that channel length scaling for the planar Si MOSFET is ending. At the end of Moore's law or MOSFET scaling there is no hard limit rather the transistor becomes increasingly less functional (useful) when scaled due to increased off-state leakage and/or little improved switching performance. Once leakage prevents further scaling of the planar MOSFET, the microelectronics revolution will still continue with innovation coming from the introduction of new materials, device structures, circuits, and architecture innovation. Several of the circuit and architectural innovations were presented in the earlier part of this book.

During this new post Moore's Law era, to continue making progress it is instructive to understand (i) the physical mechanisms responsible for the off-state leakage, (ii) what MOSFET features are currently implemented to control leakage and (iii) the prospect of new device structures based on an off-state leakage metric. All of which are included in this chapter.

Chapter 1 provided an overview of different leakage components. Section 12.2 to Section 12.6 in this chapter covers the dominate source of leakage in nanoscale transistors, why it occurs and state-of-the-art transistor features used to reduce leakage in 35 nm and 45 nm logic transistors. Many of these or similar concepts will be necessary in any device technology to compete with the industry standard, planar Si MOSFET. In Section 12.7 future solutions for transistor leakage using materials and structures are discussed.

12.2 SUB-THRESHOLD LEAKAGE IN NANOSCALE PLANAR SI MOSFETS

Today's planar Si MOSFET is far away from any thermodynamic or quantum mechanical limits, yet source to drain leakage is at the practical limit of ~100nA/um since the MOSFET is approaching its minimum channel length limit (L_E.) In this section we describe the dominate source of off-state leakage which in a well designed MOSFET is the source-to-drain sub-threshold leakage and why this leakage sets the ultimate minimum channel length possible for a planar MOSFET.

There are many sources of off-state leakage in a start-of-the-art MOSFET which have been covered previously. There have been several recent studies as to which leakage mechanism dominates near the approximately 20 nm planar CMOS channel length limit. The general consensus is that sub-threshold leakage is dominant over other types of leakage like band-to-band tunneling. The sub-threshold leakage results from the close proximity of the drain to the source which causes the source energy barrier height to be reduced by drain induced barrier lowering (DIBL) as described in Chapter 1. Sub-threshold leakage then results from increased thermal emission of carriers over the reduced energy barrier. Drain electrostatics now limit significant further scaling of the planar MOSFET due to the nanoscale source to drain spacing which has decreased from 100 um to ~20 nm in order to satisfy Moore's Law which requires the source to drain distance to decrease by $\sqrt{2}$ every two years to support a doubling of the transistor density.

Historically, to control DIBL, a simple relationship between the channel length and oxide thickness (t_{ox}) has been followed which has a simple physical origin. To maintain low off-state leakage and robust gate control over the source barrier height, the channel length was maintained at $L_E \geq 20t_{ox}$. The relationship between oxide thickness, channel length, and small drain induced barrier lowering results due to requirements on (i) the shape of the depletion layer of a MOSFET and (ii) the sub-threshold slope of the MOSFET needs to be <80mV/decade.

For the first requirement, the depletion layer under the gate can be approximated by a rectangle. The magnitude of drain-induced barrier lowering is set by the aspect ratio of the rectangle and requires the length of the rectangle (channel length) to be 2X the width (depletion layer depth).

The second requirement adequately small (~80 mV/decade) sub-threshold slope (S) is needed to maintain a large on-current to off-current ratio. The sub-threshold slope is a measure of the gate-voltage below threshold required to reduce the off-state leakage (10X) and can be expressed as:

$$S = (1 + \frac{\varepsilon_{si} t_{ox}}{\varepsilon_{ox} W_{dm}})(\ln 10)\frac{kT}{q} \qquad (1)$$

Where W_{dm} is the depletion layer depth, and ε_{Si} and ε_{ox} are the permittivity of the oxide and silicon. From Eq. (1), the ideal value of S is 60 mV/decade and occurs when $t_{ox} \ll W_{dm}$. A reasonable upper limit on S is ~ 80 mV/decade which requires

$$t_{ox} / W_{dm} \approx 0.1 \qquad (2)$$

From Eq. (1) thus a limit on channel length that needs to be at least twice W_{dm} is given by,

$$L_E \geq 20 t_{ox} \qquad (3)$$

For silicon dioxide gate dielectric, the practical minimum thickness due to tunneling currents (required to be approximately one-tenth or less than the sub-threshold leakage due to design performance constraints) is approximately 1.2 nm. This then requires the minimum MOSFET channel length limit to be ~24 nm which is close to the physical 35 nm gate in state of the art production for the 65 nm node. The limit suggested in (3) provides a lower bound. Empirical data shows that for good electrostatics L_E has to be at least $40t_{ox}$.

In the next sub-section we present a simple model for sub-threshold leakage to have a better understanding of the leakage mechanism before talking about the technologies to control this leakage in sub 100nm devices.

12.2.1 Sub-threshold Leakage Model

In order to appreciate the impact of sub-threshold leakage in state-of-the-art devices it is necessary to have a good understanding of the underlying physical mechanism. In this section we provide a simple model which outlines the leakage phenomenon. The often neglected issue with sub-threshold leakage is that most of the leakage does not result from the standard design rule structure, and all device layouts do not have the same leakage. Sub-threshold leakage is exponentially dependent on the threshold voltage and is given by the following relationship,

$$I_{exp} = \mu_0 C_{ox} \frac{W_{eff} V_t^2}{L_{eff}} e^{1.8} e^{[(V_{gs} - V_{th})/(nV_t)]} \left[1 - e^{-V_{ds}/V_t}\right] \qquad (4)$$

where, V_{ds} and V_{gs} are the voltages from the drain and gate to the source, V_{th} is the threshold voltage and V_t is the thermal voltage (kT). W_{eff} and L_{eff} are the effective gate width and length, μ_0 is the mobility which depends on the temperature and doping profile, C_{ox} is the gate capacitance per unit area and n is the sub-threshold swing coefficient. From the above equation threshold voltage in sub-100 nm technology is not constant and depends on effective channel length (L_{eff}) and width (W_{eff}). This requires remodeling of the threshold voltage in order to make accurate calculations of speed and power consumption.

A simple one-dimensional second order threshold voltage model can be used to describe the various short channel effects. Short channel effects decrease the threshold voltage due to 2-D electrostatic charge sharing between the gate and source-drain extensions. To obtain the short channel model of the threshold voltage from the long channel model, three kinds of short channel effects need to be included:

- $\Delta V_{th} (L_{eff})$: caused due to channel length modulation.
- $\Delta V_{th} (W)$: caused due to the field oxide charge or from the parasitic side wall device.
- $\Delta V_{th} (DIBL)$: where DIBL refers to the *drain induced barrier lowering*.

$$V_{th} = V_{thL} - \Delta V_{th}(L_{eff}) + \Delta V_{th}(W) - \Delta V_{th}(DIBL) \qquad (5)$$

The V_{thL} in (5) is the long channel threshold voltage.

In order to keep the above mentioned 2D electrostatic short channel effects on threshold voltage under control, the gate oxide thickness is reduced nearly in proportion to the channel length. This is necessary in order for the gate to retain more control over the channel than the drain. It has been observed empirically that for optimum performance without wasting any substrate wafer area, the effective channel length must be approximately at least 40 times the oxide thickness.

After understanding the physical mechanism of sub-threshold leakage we next look at one of the more important device features to control sub-threshold leakage, namely angled implants implemented in nearly all advanced technologies.

12.2.2 Large-angled Implants

The most important device change during the past decade to control sub-threshold current was the addition of large angle well implants called "halos". The *halo-implant* is a high angle implant of well type dopant species introduced into the device after transistor gate patterning (for example, n-type dopant implanted into the n-well of a pMOSFET). Typically the same lithography step as the source/drain extension implant is used. Because of the high halo implant angle, multiple implants with rotations are needed to ensure uniform doping on all sides of the channel.

During the last decade, two types of halo implant profiles have been targeted. The first type, called a super-halo, is a highly non-uniformly doped pocket implant used to reduce the source/drain extension to well depletion region for improved short channel control. The second type of halo implant is a non-localized implant designed to boost the well doping for sub-design rule gate lengths. To achieve this, the halo implants are targeted such that the lateral implant is approximately

$$Lateral = R_P \sin(\theta) \approx \frac{1}{2} L_{GATE} \tag{6}$$

Where R_P is the implant projected range, θ is the target angle and L_{GATE} is the gate length. In the boosted well technology, instead of having two separate halo regions as for the super halo, the sole and significant advantage of well boosting implants is the increased channel dopant concentration created in sub-design rule gate lengths. Higher well doping in sub-design rule structures is useful to compensate for the increased leakage which results from increased drain induced barrier lowering. It is the significantly higher sub-threshold leakage in sub-design rule structures, which are always present due to variations in the critical dimensions that dominate the total leakage standby power. The higher well doping in the sub-design rule gate structures creates a much flatter off-state leakage vs. gate length.

The well boosting halo implants were added during the 1990's to slow the increase in sub-threshold leakage. The sub-threshold leakage increased rapidly during the past decade driven by rapid power supply and channel length scaling. During the early years in the microelectronic industry little power supply scaling occurred. However in the last decade to control active power, the supply voltage has been scaled close to the practical high performance limit of 1.0 V. Supply voltage scaling also affects sub-threshold leakage since in order to maintain adequate gate overdrive, the MOSFET threshold voltage needs to be reduced. This directly leads to the exponential rise of sub-threshold current for smaller channel lengths.

12.3 SION DIELECTRICS TO REDUCE GATE TO CHANNEL DIRECT TUNNELING CURRENT

The desire to improve device performance and reduce the predominant sub-threshold leakage currents has not only led to small channel lengths but also the aggressive scaling of the gate oxide thickness to below 2nm which brings it to the direct tunneling (DT) regime. It has also been shown that not only electron but hole tunneling becomes predominant at such dimensions. The leakage due to DT is measured per unit area, and a certain criteria of $1A/cm^2$ was thought to be the ultimate limit of scalable oxide thickness, but today's state of the art technology can sustain gate leakage currents of the order of $100A/cm^2$ under the assumption that only 1% of the total chip area is taken up for making the gates. As the gate oxide is scaled below 20 gate leakage is predicted to increase at a rate of more than 500X per technology generation, while sub-threshold leakage increases by around 5X per technology generation. This sets a lower limit for gate oxide thickness in the 1-1.5nm range.

As a result there has been an immense interest in alternative gate dielectrics with higher relative permittivity facilitating the use of thicker dielectric films. Silicon nitride was the obvious choice due its compatibility with the existing CMOS process. High quality SiN films exhibit relatively low densities of interface traps, fixed charge, and bulk traps. Gate leakage current in silicon nitride is significantly lower than a silicon dioxide film of the same equivalent oxide thickness. Also silicon nitride films exhibit very strong resistance to boron penetration and oxidation at high temperatures. These properties, coupled with its room temperature deposition process, have made it an attractive candidate to succeed thermal silicon dioxide as an advanced gate dielectric in future generations of ULSI devices. Other high-K dielectrics though at first glance are expected to give better tunneling leakage performance due to the thicker film for a given equivalent oxide thickness are in fact far worse. This is more elaborately discussed in Section 12.7.1.1.

12.4 OFFSET SPACERS TO REDUCE EDGE DIRECT TUNNELING CURRENT

As stated previously, off-state leakage is comprised of gate and sub-threshold leakage. Edge direct tunneling (EDT) current is often the dominate gate leakage mechanism in scaled CMOS with less than 2 nm gate-oxide thickness. Edge direct tunneling leakage current has become a critical issue in CMOS scaling since gate-to-SDE overlap area (as a percentage of gate area) increases with every technology generation (almost half the channel length is covered by SDE region at the 65 nm and 90 nm nodes). In order to control the gate overlap area and reduce the gate capacitance, *offset spacers* are sometime used in the present day sub-100 nm CMOS technology. By varying the thickness of the offset spacer, the gate-to-SDE overlap area and the junction depth can be independently varied. The offset spacer elongates the effective channel length and improves the roll-off characteristics. Also as the effective gate-to-SDE overlap area reduces, so does the EDT off-state leakage current.

There exists an optimum spacer width which is set so as to achieve the highest drive current with tradeoffs in short channel effects, external resistance and gate-to-SDE coupling. It has been shown that a minimum gate to SDE overlap of 15-20 nm is required to prevent drive current (I_{Dsat}) degradation when the gate oxide thickness is 4.5 nm for process flow similar to that of a 0.25 μm CMOS technology. It was also shown that scaling of SDE vertical depths below 30-40 nm showed little or no performance benefit for 0.1um devices and beyond for, any improvement in short channel effects due to reduced charge sharing is offset by a large increase in external resistance and poor gate coupling between the channel and the extensions.

12.5 COMPENSATION IMPLANTS TO REDUCE JUNCTION LEAKAGE

In the past decade major concentration in leakage considerations was given to reducing sub-threshold and gate leakage currents. This was achieved by high doping concentration in the channel region (halo implants etc. to reduce short channel effects). The proximity of the valence and conduction bands in the depletion region of the junctions as a result of the high concentrations produced a tunneling current which has started becoming important with aggressive scaling over the years. In order to reduce this band-to-band tunneling in the bulk junction due to high dopant concentration, *compensation implants* are used. Compensation implants are introduced into both NMOS and PMOS devices in the same lithography sequence used for source/drain implants. This implant uses the same type species as the source/drain implant but with a lower dose and higher energy to give a more graded implant profile at the junction. The compensation implant is done in such a way so as to give a 20-30% reduction in the junction capacitance with no degradation in the isolation performance or the implant penetration of the gate oxide.

Even though junction leakage is high (of the order of 1 nA/um) for a gate length of 30 nm, it is orders of magnitude lesser than the corresponding values for gate, sub-threshold and gate-induced drain leakage. For even shorter channel length devices, and assuming a 1.6X doping concentration increase per technology generation, the junction leakage current is still far below a value of 1uA/um, the upper leakage limit. Hence we see that even in future ULSI device technologies junction leakage is not of much concern if properly designed compensation implants are used.

The next section provides a simple model for the junction leakage current which is similar to the leakage models existing for estimating tunneling probability in zener diodes.

12.5.1 Equivalent Diode Tunneling Model for Junction Leakage

Junction leakage arises from the high doping concentration in the channel region required to attain threshold voltages, and to limit short channel effects in aggressively scaled devices. With the increase in the dopant concentration, the width of the depletion region at a given reverse bias decreases, and the energy bands in the depletion region bend more steeply. Because of the wave nature of the electrons, there is a finite probability that valence electrons in a p-type semiconductor tunnel through the forbidden region and appear at the same energy at the conduction band. As the temperature increases there will be a significant increase in leakage current due to the increased influx and kinetic energy of valence band electrons available for tunneling at that temperature. A simple equivalent diode tunneling model can be used to calculate the tunneling probability for high doping concentrations.

The following is such a simple tunneling model used for calculating the tunneling probability similar to the one used for zener diodes. The probability of tunneling Θ can be approximated using the barrier height in the WKB approximation equation:

$$\Theta = \exp\left(\frac{B}{V}\right) = \exp\left(\frac{-qBL}{E_g}\right) \tag{7}$$

$$B = \frac{4\sqrt{2m^*}E_g^{3/2}}{3q\hbar} \tag{8}$$

Where E_g is the bandgap energy, m* is the effective mass, q is the electric charge, and ħ is the Planck's constant. The tunneling current I can be written as:

$$I = qANv\Theta \tag{9}$$

Where A is the area, N is the density of electrons in the valence area, and v is the electron velocity.

12.6 SOURCE/DRAIN EXTENSION GRADING TO REDUCE GATE INDUCED DRAIN LEAKAGE (GIDL)

As the CMOS is scaled below the 100 nm regime another mechanism which imposes a limit on the power supply and gate oxide thickness is the gate-induced drain leakage (GIDL) current. GIDL is caused due to the band-to-band tunneling in the drain region underneath the gate. A large gate-to-drain bias can cause sufficient band bending near the interface between silicon and the gate dielectric for the valence band electrons to tunnel into the conduction band. GIDL becomes less significant for digital applications as the power supply is scaled to about 1V (close to the silicon band gap), but it is still important in memory applications (DRAM) where data retention is severely degraded by GIDL current.

Significant gate-induced drain leakage current can be detected in thin gate oxide MOSFETs at drain voltages much lower than the junction breakdown voltage. It is established from the GIDL model described in the next section that the electric field across the tunneling barrier is very sensitive to the drain doping gradient. With this in mind, visible means to reduce GIDL is the use of Lightly Doped Drain (LDD) CMOS devices with a graded drain region. It is desirable to limit the GIDL current to 0.1 pA/um. For this the field oxide in the drain overlap region must be limited to 1.9 MV/cm. This sets a limit on the oxide thickness and supply voltage as follows:

$$V_{cc} = 1.2 + t_{ox} \times 1.9 MV / cm \tag{10}$$

Full-overlap LDD devices are used to achieve this because the lateral field is suppressed while the drain concentration is high enough so that the dominant tunneling point has a band bending of 1.2 eV. In order to achieve this, a minimum n- doping of $10^{18} cm^{-3}$ is required to raise the V_{dg} over that of a non-LDD MOS. LDD MOS usually have a moderately doped n- region in order to reduce the series resistance and minimize the hot-electron degradation. If the doping concentration is higher than 10^{18}cm^{-3} at the gate edge, a point with doping 10^{18}cm^{-3} always exists in the gate/drain overlap region, which makes LDD MOSFETs no better than conventional MOSFETs for the GIDL. Buried LDD MOSFETs with a peak n- doping concentration several hundred angstroms underneath the Si-SiO2 interface seems to be the more favorable device structure for both hot-electron reliability and gate-induced drain leakage current.

Another scheme is to use a graded gate oxide structure to reduce the leakage current since the T_{ox} above the point where the doping concentration is 10^{18}cm^{-3} comes into picture. Judicious design of the gate edge profile to compromise device reliability and gate-induced drain leakage current is necessary if the graded gate-oxide structure is adopted. In all, LDD devices will play a major role in future ULSI technology in suppressing GIDL. The next section presents a simple model to calculate the GIDL current.

12.6.1 A Simple GIDL Model

Gate-induced drain leakage current is found to be due to the band-to-band tunneling occurring in the deep-depletion layer in the gate-to-drain overlap region. When high voltage is applied to the drain with the gate grounded, a deep-depletion region is formed underneath the gate-to-drain overlap region. Electron-hole pairs are generated by the tunneling of valence band electrons into the conduction band and collected by the drain and the substrate, separately. Since all the minority carriers generated thermally or by band to band tunneling in the drain region flow to the substrate due to the lateral field, the deep depletion region is always present and the band-to-band tunneling process can continue without creating an inversion layer.

Band-to-band tunneling is only possible in the presence of a high electric field and when the band bending is larger than the energy gap, E_g. The field in silicon at the Si-SiO2 interface also depends on the doping concentration in the diffusion region and the difference between V_D and V_G, i.e. V_{DG}. A simple expression for the surface field at the dominant tunneling point can be expressed as:

$$E_s \approx \frac{V_{gd} - 1.2}{3t_{ox}} \tag{11}$$

Where E_s is the vertical electric field at the silicon surface, 3 is the ratio of silicon and oxide permittivity, and t_{ox} is the thickness of the oxide in the overlap region. Band-to-band tunneling current density is the highest where the electric field is the largest. The theory of tunneling current predicts:

$$I_{gidl} = AE_s \exp\left(\frac{-B}{E_s}\right) \tag{12}$$

Where A is a pre-exponential constant and B = 21.3 MV/cm.

From the above equations it is clear that a band bending 1.2 eV is the minimum necessary for band-to-band tunneling to occur. This is a simple one-dimensional (1-D) band-to-band tunneling current model. From this model it is evident that the gate induced drain leakage current depends heavily on the gate oxide thickness and the impurity concentration in the gate-to-drain overlap region.

12.7 FUTURE SOLUTIONS

12.7.1 New Materials

In addition to new materials required for MOSFET structural changes like strained Si and SOI, new materials are also needed to replace existing materials in the MOSFET. Two key areas of focus for the industry are high-k gates and the self-aligned silicide (salicide), which will be discussed in this section. A suitable replacement of the silicon dioxide gate dielectric with a high-k material is perhaps the most needed and difficult project facing the US\$300 billion semiconductor industry. To put this problem in perspective, in most semiconductor systems no good insulator has ever been found making the fabrication of high performance MOSFETs impossible in nearly all-material systems except silicon. Whether nature has blessed silicon with a second insulator possessing a good interface match is still unknown. Compared to high-k gates, changing salicide materials is a much simpler problem though still difficult and has had some recent success with the evolution from TiSi2 to CoSi2 and now to NiSi [1].

12.7.1.1 High-K Gates

The task of inserting high-k gates into an advanced 45 nm logic technology [2] to achieve a net performance gain should not be overestimated. Tremendous progress has been made during the last decade on high-k gate dielectrics and sub-100 nm transistors have been demonstrated. However, at present none of the published high-k gate transistors offer performance improvement over state-of-the-art 45 nm [2] or 35 nm [3] transistors currently in high volume production. There have recently been many excellent high-k gate publications focusing on various aspects of the technology [4, 5]. The purpose of this section is to highlight many of the key challenges with high-k gate dielectrics so they can be addressed.

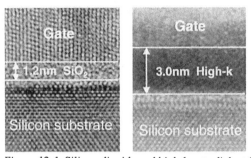

Figure 12-1. Silicon dioxide and high-k gate dielectrics.

The demonstration of a modern Si MOSFET was made possible by the SiO_2 passivation breakthrough in 1958 [6, 7] but it still took 10 more years for the Si-MOSFET to become viable and required solving technical problems like sodium ion drift [8] and hydrogen anneal passivation [9]. Introducing a high-k dielectric into the Si MOSFET appears equally or more difficult than the task of introducing SiO_2 in the 1960s. The issue with high-k films is not in the fabrication as shown by the TEM micrographs in Figure 12-1 of a gate stack with 1.2 nm silicon dioxide and physically thicker but electrically thinner high-k gate [5, 10]. The issue is the numerous requirements for a silicon dioxide replacement. Several of the key requirements are improved leakage current at the same or lower equivalent capacitance, equivalent channel mobility, equivalent or improved time to dielectric breakdown, low interface state density, low charge trapping, and compatibility with standard CMOS processing and thermal cycles. To appreciate the difficulty of the task, we will look at a few of these problems in more detail.

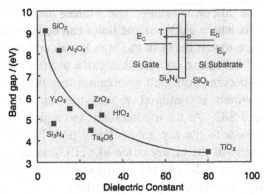

Figure 12-2. Bandgap versus dielectric constant for insulators.

First, for a high-k film to replace SiO$_2$, it must have lower gate tunneling current, which is the sole motivation for the material change. High-k films though physically thicker do not always have reduced leakage since the bandgap for most films is smaller than SiO$_2$. Figure 12-2 shows the relationship between bandgap and dielectric constant [10]. In addition to the increased leakage from the smaller bandgap, the high-k gate insulator to silicon barrier height must be larger than the supply voltage for acceptable leakage.

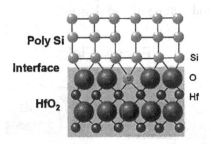

Figure 12-3. Interface state in HfO$_2$ / Si system.

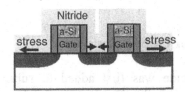

Figure 12-4. Poly-Si Gate stress memorization using S/D anneal.

Another problem for high-k films is charge trapping that shifts the threshold voltage. Commonly an ultra-thin SiO$_2$ layer is used between the high-k film and the silicon substrate. This creates another interface region with dangling bonds where electrons and holes can be traps. Still yet another issue is the interface match between the high-k film and the top and bottom gate. It is not practical to use an ultra-thin oxide at both the top and bottom interfaces since it becomes difficult to obtain a less than ~1.2 nm physical oxide thickness which is required to provide an advantage over the mainstream nitrided SiO_2. At the top interface, the high-k film needs to have low interface states with the top gate (ideally poly-Si). This is a difficult requirement for many of the binary oxide like HfO_2, since excess Si bonded to Hf as shown in Figure 12-3 can potentially create a deep level.

A high quality interface requires less than 1% of the bonds to be incorrectly bonded. Because of this very difficult requirement at the top interface, many researchers are primarily focusing on metal top gates with high-k. The integration issues with switching to metal gates should not be underestimated. For example, performance features like poly-Si gate strain created with the high stress capping layers before the source and drain anneal (Figure 12-4) will need to be equivalently replicated in a metal gates flow otherwise a transistor with a metal top gate and high-k gate dielectric will not out perform a state-of-the-art transistor. Lastly there are also geometric MOSFET structural challenges with high-k gates that cause the fringe capacitance to be much larger (Figure 12-5) which can lead to degraded short channel effects. The fringe field will place serious constraints on the extendibility and scaling of high-k gates.

Fringe Field

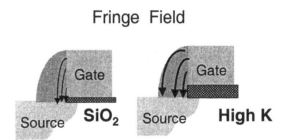

Figure 12-5. Fringe electric field for transistor with SiO2 and high-k gate dielectric.

12.7.1.2 Materials for Contact Resistance

Historically, salicide was first added to sub-micron transistors since contacts directly on heavily doped n-type and p-type source and drains started to limit device performance due to high external resistance. In the 1980's, salicide was added to reduce the contact resistance since this allows the entire source and drain to contact a metal layer. The contact metals used by the industry for 1 µm to 0.1 µm technologies have been $TiSi_2$ and $CoSi_2$. These two salicides were chosen based on their low resistivities and ability to withstand process temperatures up to 800°C without agglomerating. With the introduction of SiGe source-drain for stained Si in the 90 nm node, to improve the contact resistance, the salicide material had to be changed yet again to NiSi. Figure 12-6 shows the salicide evolution during the past decade. This evolution of salicide contact metals resulted for several reasons.

The dominant considerations for choosing a contact metal are sheet resistance, diode leakage on shallow junctions, silicide to silicon interface resistance, and maintaining low sheet resistance on small geometries. The industry in the 1980s and 90s extensively used titanium salicide. The key-scaling problem with $TiSi_2$ is the difficulty of forming the low resistivity C54 phase on narrow lines. The nucleation and growth of the high resistivity C49 phase forms first and it is very difficult to transform this to the low resistivity phase on narrow lines. As a result $CoSi_2$ was widely adopted by the industry at the 180 nm technology node since it forms well on narrow lines. In fact cobalt salicide scales well to 45 nm feature size [1] however it does not form well on SiGe. Like the issue with $TiSi_2$ on narrow lines, Ge in the source and drain regions inhibits $CoSi_2$ transition to the low resistivity disilicide phase. To overcome this problem for $CoSi_2$, it is necessary to add an additional sacrificial Si buffer layer in the source and drain but this adds additional cost and complexity. NiSi can support low resistance silicides directly on Ge as shown previously [1, 11].

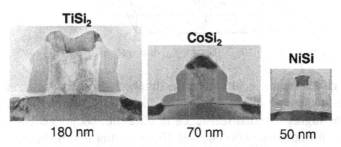

Figure 12-6. Salicide technology evolution.

It is expected that as the industry moves to strained Si, NiSi will be widely adopted. As technology scales, there will be continued need for new materials that can improve the extrinsic resistance are narrow bandgap semiconductors in the source and drain. Without such innovations the leakage current required to achieve the target performance will increase further.

12.7.2 New Structures

Historical transistor improvements (expect the recent introduction of strained Si or SOI) have come from channel length scaling, which improves the intrinsic MOSFET switch. Thus the majority of future transistor work has focused on making a better intrinsic switch [10, 12, 13, 14, 15, 16]. Common novel structures with new materials to improve intrinsic devices performance include ultra-thin body devices and carbon nanotubes.

12.7.2.1 Ultra-thin Body MOSFET

The key device concept behind ultra-thin bodies is improved short channel effects. Ultra-thin bodies include double gate or tri-gate MOSFETs as shown in Figure 12-7. In order to achieve good short channel effects in these devices, the body thickness needs to be considerably thinner than the gate length [17, 18]. This causes two issues: high extrinsic source and drain resistance and a requirement that the body be undoped.

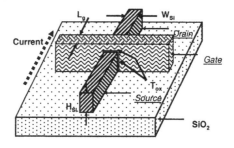

Figure 12-7. FINFET ($H_{Si} \gg W_{Si}$) or tri-gate transistor ($H_{Si} \sim W_{Si}$).

The high resistance results since the thin body is difficult to contact even with state-of-the-art process techniques like raised source and drains. At present, the higher external resistance in ultra-thin bodies generally negates the benefit of improved short channel effect. Ultra thin bodies are required to be undoped to maintain acceptable dopant fluctuation driven threshold voltage variation. Undoped body devices require a complicated process flow of dual metal gates to set the threshold voltages.

Before ultra-thin body devices can be viable, breakthroughs in contact resistance and undoped body fabrication are needed. However, perhaps the biggest challenge facing ultra thin body devices is the compatibility with current state-of-the-art performance enhancement concepts. For a new device structure or material to make it into production, the integrated product needs to be improved.

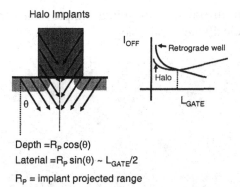

Depth $= R_P \cos(\theta)$

Laterial $= R_P \sin(\theta) \sim L_{GATE}/2$

$R_P =$ implant projected range

Figure 12-8. Halo implants used to "flatten" transistor off-state leakage vs. gate length.

At present some of the performance enhancement concepts already in production that provide greater than 30% performance boost like high angle halo implants and uniaxial stress have not and can not be implemented into ultra-thin body devices. To further make this point, today all high performance sub 100 nm technologies use high angle halo implants to engineer the shape of the off-state leakage vs. gate length (see Figure 12-8). To benchmark off-state leakage correctly, the performance comparison is not a single transistor at a fixed off-state leakage but rather a chip at a fixed chip standby-leakage where the leakage is dominated by leakage from sub-design rule transistors always present due to gate length variation. Halo implants improve product performance by over 10% and for ultra thin body devices to compete similar enhancements concepts are needed in the new device structure.

12.7.2.2 Carbon Nanotubes

Carbon nanotubes are another new structure and material that has received much focus in recent years for both active devices and interconnects. Carbon nanotubes have potential due to many advantageous properties: 1-dimensional current transport, yield strength, and nanoscale diameter, which are controlled by chemistry and not fabrication. A SEM micrograph of carbon nanotubes is shown in Figure 12-9 [19]. The field of carbon nanotubes has too numerous challenges today to list and its long term potential is still unclear if this technology can ever replace Si MOSFETs. At present nanoscale gate length carbon nanotubes, just like Si MOSFETs, suffer from high off-state leakage caused by ambipolar thin gate oxide induced tunneling currents and will likely also suffer from source-to-drain tunneling leakage currents caused by the low electron and hole conductivity effective masses. Hence at the nanoscale device limit, carbon nanotubes transistors may offer some advantage over Si MOSFETs but it appears to be much smaller than an order of magnitude improvement often required in implementing a radically new technology.

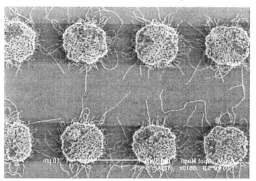

Figure 12-9. Carbon nanotube structures.

12.8 SUMMARY

Conventional planar bulk MOSFET channel length scaling, which has driven the industry for the last 40 years, is slowing. To continue Moore's law into the nanometer regime innovative solutions are needed to tackle the challenge of high leakage currents. Large angled well boosting implants to reduce sub-threshold leakage, offset spacers to reduce edge-direct tunneling, compensation implants to reduce junction leakage and a lightly-doped-drain structure to reduce GIDL are some of the steps in this direction.

Novel materials (high-K dielectrics, metal gates) that address MOSFET poly-Si gate depletion, gate thickness scaling and alternate device structures (FinFET, tri-gate, or carbon nanotube) are possible technology directions for future CMOS technology generations. Without these enhancements the intrinsic device electrostatics and extrinsic device resistance will be poor, resulting in worse switching behavior and a faster increase in leakage currents. These enhancements along with circuit and architectural solutions discussed in the earlier chapters will be essential to deal with leakage currents in nanometer scale transistor switch structures.

REFERENCES

[1]   S. Thompson and et al. "A 90nm Logic Technology Featuring 50nm Strained Silicon Channel Transistors, 7 layers of Cu Interconnects, Low k ILD, and 1 mm^2 SRAM Cell," *Technical Digest of the IEEE International Electron Devices Meeting*, Washington, 2002, pp 61-64.

[2]   S. E. Thompson and et al., "A Logic Nanotechnology Featuring Strained Silicon," *IEEE Electron Device Letter*, vol. 25, pp. 191-193, 2004.

[3] P. Bai, "A 65nm Logic Technology Featuring 35nm Gate Lenghts, Enhanced Channel Strain, 8 cu interconnects Layesrs, Low-k ILD and 0.57um2 SRAM Cell," *Technical Digest of the IEEE International Electron Devices Meeting*, vol. pp., 2004.

[4] Z. Ren, M. V. Fischetti, E. P. Gusev, E. A. Cartier, and M. Chudzik, "Inversion channel mobility in high-[kappa] high performance MOSFETs," *Technical Digest of the IEEE International Electron Devices Meeting*, San Francisco, 2003, pp 33.2.1-4.

[5] D. Barlage, R. Arghavani, G. Dewey, M. Doczy, B. Doyle, J. Kavalieros, A. Murthy, B. Roberds, P. Stokley, and R. Chau, "High-frequency response of 100 nm integrated CMOS transistors with high-K gate dielectrics," *International Electron Devices Meeting. Technical Digest*, pp. 10.6.1-4, 2001.

[6] D. Kahng and M. M. Atalla. Silicon-silicon dioxide field induced surface devices. IRE-AIEE Solid-State Device Research Conference, Carnegie Institute of Technology, Pittsburgh, PA, 1960.

[7] M. M. Atalla, M. Tannenbaum and E. J. Scheibner, "Stabilization of silicon surface by thermally grown oxides," *Bell Syst. Tech. J.*, vol. 38, pp. 123, 1959.

[8] E. H. Snow, B. E. Deal, A. S. Grove and C.-T. Sah, "Ion transport phenomena in insulating films using the MOS structure," *J. Appl. Phys.*, vol. 36, pp. 1664-1673, 1965.

[9] P. Balk, "Effects of hydrogen annealing on silicon surfaces," *Electrochemical Society Spring Meeting*, San Francisco, CA, 1965.

[10] B. Doyle, R. Arghavani, D. Barlage, S. Datta, S. Doczy, J. Kavalieros, A. Murthy and R. Chau, "Transistor elements for 30nm physical gate lengths and beyond," *Intel Technology Journal*, vol. pp., 2002.

[11] T. Ghani, et. al., "A 90nm High Volume Manufacturing Logic Technology Featuring Novel 45nm Gate Length Strained Silicon CMOS Transistors," *Technical Digest of the IEEE International Electron Devices Meeting*, San Francisco, 2003, pp 978-980.

[12] Y. Bin, C. Leland, S. Ahmed, W. Haihong, S. Bell, Y. Chih-Yuh, C. Tabery, H. Chau, X. Qi, K. Tsu-Jae, J. Bokor, H. Chenming, L. Ming-Ren and D. Kyser, "FinFET scaling to 10 nm gate length," *International Electron Devices Meeting. Technical Digest*, pp. 251-254, 2002.

[13] R. Chau, B. Boyanov, B. Doyle, M. Doczy, S. Datta, S. Hareland, B. Jin, J. Kavalieros and M. Metz, "Silicon nano-transistors for logic applications," *Physica E: Low-dimensional Systems and Nanostructures*, vol. 19, pp. 1-5, 2003.

[14] B. S. Doyle, S. Datta, M. Doczy, S. Hareland, B. Jin, J. Kavalieros, T. Linton, A. Murthy, R. Rios and a. Chau et, "High performance fully-depleted tri-gate CMOS transistors," *IEEE Electron Device Letters*, vol. 24, pp. 263-265, 2003.

[15] R. Chau, J. Kavalieros, B. Doyle, A. Murthy, N. Paulsen, D. Lionberger, D. Barlage, R. Arghavani, B. Roberds and a. Doczy et, "A 50 nm depleted-substrate CMOS transistor (DST)," *International Electron Devices Meeting. Technical Digest*, pp. 29.1.1-4, 2001.

[16] R. Chau, J. Kavalieros, B. Roberds, R. Schenker, D. Lionberger, D. Barlage, B. Doyle, R. Arghavani, A. Murthy and a. Dewey et, "30 nm physical gate length CMOS transistors with 1.0 ps n-MOS and 1.7 ps p-MOS gate delays," *International Electron Devices Meeting 2000*, pp. 45-48, 2000.

[17] J. G. Fossum, M. M. Chowdhury, V. P. Trivedi, T.-J. King, Y.-K. Choi, J. An and B. Yu, "Physical insights on design and modeling of nanoscale FinFETs," *International Electron Devices Meeting*, pp. 679-682, 2003.

[18] V. P. Trivedi and J. G. Fossum, "Scaling fully depleted SOI CMOS," *IEEE Trans. Electron Devices*, vol. 50, pp. 2095-2103, 2003.

[19] A. Ural, L. Yiming and D. Hongjie, "Electric-field-aligned growth of single-walled carbon nanotubes on surfaces," *Applied Physics Letters*, vol. 81, pp. 3464-3466, 2002.

INDEX